Cybersecurity Lessons from CoVID-19

Cybersecurity Lessons from CoVID-19

Robert Slade

CRC Press
Taylor & Francis Group
Boca Raton London New York

CRC Press is an imprint of the
Taylor & Francis Group, an **informa** business

First edition published 2021
by CRC Press
6000 Broken Sound Parkway NW, Suite 300, Boca Raton, FL 33487–2742

and by CRC Press
2 Park Square, Milton Park, Abingdon, Oxon, OX14 4RN

ISBN: 978-0-367-68269-9 (hbk)
ISBN: 978-1-003-13667-5 (ebk)

Typeset in Times
by SPi Global, India

To Gloria

Contents

Author		ix
Introduction		xi
1	Expanding Security: The CIA Triad	1
2	Security Management	17
3	Access Control	31
4	Security Architecture	35
5	Business Continuity and Disaster Recovery Planning	41
6	Physical Security	51
7	Cryptography (Yes, Really)	55
8	Application Security	57
9	Security Operations	67
10	Telecommunications and Networking Security	79
11	Law, Investigation, and Ethics	93
12	Be Kind. Be Calm. Be Safe	97
13	Will We Win?	103
Index		107

Author

Robert Slade worked his way through university in a hospital and later as an industrial first aid attendant. Later he found out he was a teacher, and, as an information security maven, he taught on six continents. Along the way, he also wrote books on computer viruses, software forensics, and security terminology. (*Robert Slade's Guide to Computer Viruses*, 1995, 1996, Springer-Verlag; *Viruses Revealed*, 2001, Osborne McGraw-Hill: *Software Forensics*, 2004, McGraw-Hill; and *Dictionary of Information Security*, 2006, Syngress). He'd much rather be spending time with his grandchildren than writing another book. It is next to impossible to get him to take 'bio' writing seriously, but more information than anyone would want to know about him is available on Twitter @rslade.

Introduction*

Since much of this book will deal with disasters and horrors and since all of you will have numerous examples of how people behaved badly during the pandemic, let me start on a positive note.

A young man from Thailand met his girlfriend in Singapore. Singapore, early on, became a CoVID-19 hotspot. The Thai family lives in Bangkok, which also reported CoVID-19 cases, although the numbers at the time weren't as high as Singapore's.

When they heard about the panic-buying and health equipment like masks and hand sanitizers being out of stock, they managed to get their hands on a variety of masks to send over. They threw in a bottle of hand sanitizer. They paid 860 Thai baht for international express shipping just so their friends could get the package as soon as possible. That roughly converts to a whopping US$30, which probably cost more than the masks and the sanitizer.

Disasters bring out the worst in people, but they also bring out the best.

Here is the first lesson about security. This is a fundamental point about security that I try to emphasize to my students all the time. People are, at one and the same time, both your greatest security weakness and your greatest security strength. People get tired, make errors, get fooled, and make bad decisions. Machines don't (if programmed properly, which hardly ever happens). But people can also figure out that someone who is following all the rules might still be trying to pull a fast one. Machines can't.

I am seeing the CoVID-19 pandemic from a position of privilege. My job didn't suddenly disappear, and my finances won't either unless the entire world economy collapses. (If it does, you aren't going to be reading this book, now are you?) In addition, I live in Canada, which has a decent medical system if I need it. (On the other hand, I am old, male, and fat and have diabetes and high blood pressure, all major risk factors for worst-case outcomes if I do get infected, so if any stray SARS-CoV-2 virus lands on me, I'm toast.)

Toilet paper? I had some, partly since I'm an emergency volunteer and it's part of the stock I tend to keep around as the household disaster kit, but mostly because I'm cheap and hate to buy it at full price, so I tend to buy a lot when it's on sale. (It doesn't go bad.) Haircut? I learned—long, long ago—that I

---

* Units of measure: this text has, deliberately, used both US and SI units of measure, particularly in regard to 'two metres' and 'six feet' of distancing separation.

am not handsome enough to get any of those jobs where initial appearance is vital, like salesman, con man, news anchor, politician, or celebrity supermodel. I keep my hair fairly short, but my preferred hairstyle is best described as 'low maintenance'.

I live in British Columbia (BC), which is a beautiful place to have to ride out a crisis if you can't travel, and it is supremely fortunate in the government that we have in office at this particular time, our provincial Health Minister, Adrian Dix, and our Chief Medical Health Officer, Dr. Bonnie Henry. (We were oddly, and randomly, lucky in that the dates of the school spring break, set possibly years in advance, just happened to fall at the right time so that parents were prepared to have the kids out of school but had not yet left for vacations and travel that could have been devastating.) On a regular basis, we have an official update (which is generally referred to as 'The Dr. Bonnie Show co-starring Adrian Dix and [ASL interpreter] Nigel Howard'). (You will, in the course of this book, hear more about Dr. Bonnie.) The material presented is clear, honest, factual, informative, as comprehensive as possible given the gaping holes in our knowledge about the virus, and, oddly, quite reassuring. In addition, the decisions and orders that have been taken have kept BC safer than just about any other jurisdiction of its size or larger. (I can, on the other hand, quite honestly say, given the vagaries of geographic outbreaks, that as of March 22, 2020, 90% of the CoVID-related deaths that had occurred in BC and 50% of those in Canada had occurred within three blocks of me. It's possible I am a dangerous person to know.)

Many people date the pandemic from March 11. This is, after all, the date on which the World Health Organization was first willing to use the word 'pandemic'. It is also the date of the infamous basketball game that basically ended sports as we know it. For me, though, the date was March 10. March is a big month for the security community in Vancouver. At coffee time on the morning of March 10, I had two conferences and two presentations lined up. By dinner time, it was all gone.

I recently saw a headline that asked, 'Have you lost your purpose since CoVID-19?' I realize that many people would: most people either are defined (by society) or define themselves by the job they hold. I've been self-employed or making my own jobs for three decades. And most of my colleagues are in jobs where they can, and are helping others to, work from home. We're busier than ever. I don't have enough hours in the day.

Mental health turns out to have been a much bigger issue during this pandemic than I would have expected. I recently attended a (virtual) meeting with some of the leaders in the security world, and the first few minutes were filled with stories of having to deal with stress, having to take time off, needing to force people to take time off, and other such issues. So I guess one of the next security lessons is that you have to take care of yourself. Which is easy to say but possibly difficult to do. When you are under stress, you make bad assessments. And one of the assessments that may be in error is how stressed you are and therefore how bad the decisions you're making are. This is related to the impairment of how well you can determine yourself to be impaired if you are impaired and also the Dunning–Kruger effect. But taking care of yourself does lead us to a couple of other points.

A friend, who is a doctor, asked whether he should volunteer to go back to work, as a front-line physician, during the coronavirus/CoVID-19/SARS-CoV-2 crisis. So what can retired doctors (and nurses and engineers, etc.) do during CoVID-19?

As we know from emergency management, the first rule is not only 'First, do no harm' but more specifically 'First, do no harm—to yourself'. You (and I) are old, Father William. Not only are the old particularly susceptible to the virus, but (for reasons not yet either verified or fully understood) front-line medical workers also seem to be particularly susceptible. Going back in as a front-line worker may not be the best use of your skills and experience.

Right off the top, may I suggest that you start reaching out to other retired doctors (and other professionals)? I know that lots of them have already volunteered to return to active service. But I suspect that some of them are concerned, as you are, about the dangers (or are under family pressure) and may be wondering about how best to proceed. You could be the founder of Doctors (Nurses, Engineers, etc.) Without Pay Cheques and get a wealth of skill and experience mobilized in some interesting new directions. Yes, with the hoarding and panic-buying going on, we are having to spend more time just getting the necessities of life, but we still have some time to contribute.

One of the first things to do is search out and fight mis- and disinformation with facts and analysis. Most people don't have your knowledge of disease, disease vectors, hygiene rules, transmission vectors, the lifetime of an infectious agent outside the body, the difference in effectiveness between soap and hand sanitizer, or even the difference between a bacterium and a virus. Most people are scared because they are relatively ignorant about these things. (They may not be completely ignorant and may not be ignorant of other things, but they don't know as much as you do about the specifics.) They are bombarded by news stories about people becoming sick and people dying. They are scared (and also don't know as much as we do about risk analysis and calculation). They feel powerless, and 'wash your hands' doesn't feel like it should be enough. People feel scared and don't have enough information, so they either don't do anything and ignore the whole situation or believe any random myth they hear because at least it seems like they are doing something.

You could also reach out to students. A lot of them have a lot of time on their hands right now. Even if they aren't fully trained, most of them have more information than the general public and can help. Students will also greatly benefit from working with you and the rest of us old codgers while we try to help in various ways, and they'll get valuable practical experience in the things medical (and nursing and engineering and computer security) schools can't teach them. They've got more energy than we do. And there are some interesting projects they can assist.

As well as the misinformation, there are other questions to be researched. Not being on the front lines, you/we may not be able to research or answer them directly. But some old physician somewhere may have an insight that might be key. These are questions like: Are/Why are front-line medical workers at greater risk of CoVID-19? Is ibuprofen actually a risk? Are vitamins C and D actually a benefit? When (and how much)? Lots of questions fall into this category.

But, probably more importantly, there are some projects that can be undertaken without direct front-line contact or exposure. For example, there was great concern over the availability of ventilators. Multiple projects are looking into low-cost, easy-to-manufacture, 'open source design' ventilators. (A number of these projects are being carried out by students.) One hospital in Canada is putting out videos showing how, with extra tubing and connectors, you can get one ventilator to support up to three patients. You need to get a retired respirologist to see if it will work and a retired engineer to see if it is feasible for manufacture by anyone with less than an engineering doctorate. Also, you, as a risk manager, need to look at the various designs, noting that (in this case) the best may very well be the enemy of the good and that something that is only 80% effective but can be built with scuba tanks and a child's toy may well save more lives than something that is 99.44% effective but requires a piece of The One True Cross to work. (You may need retired politicians to place these ideas in front of the authorities or retired businessmen to set up manufacturing.)

Think outside the box. A lot of us started work before there was a box.

Now go wash your hands.

What is the great lesson that CoVID-19 is teaching us? Well, history tends to teach us that history teaches us nothing, but let's be a bit more hopeful for a moment. There are many lessons to be learned. This book started as a compilation of lessons for my security colleagues. But a few of them are more generally appropriate. And there are a number of them.

If you believe everything you read, you had better not read. I know. Terrible thing to say. But it is true that a little knowledge is a dangerous thing. Not half so dangerous as a **lot** of ignorance, I suppose. And there have been **huge** amounts of mis- (and even dis-) information around this virus and the related disease. So, please: wash your hands. Don't drink bleach. Keep physically distant from people and don't try to find loopholes around those rules. Stick to the basics.

More problems are caused by our overreactions than by the original event. Having huge amounts of toilet paper will not keep you from getting infected.

And I'm glad you're all trying baking, but weren't you all on gluten-free diets a month ago?

Life is complex. Beware of 'simple' answers. As H. L. Mencken noted, for every complex problem, there is an answer that is simple and neat. And wrong.

Never challenge 'worse'. Never say 'things can't possibly get any worse'. Just ask those in Nova Scotia. Pandemic. Flooding. Worst mass murder in Canadian history. Helicopter crash. Death of a Snowbird. Forest fire. (Yes, we do heave a sigh and a wish for you.)

In terms of things getting worse, there is also Beirut, Lebanon. Already facing a disastrous financial crisis and endemic corruption when the pandemic hit, they also had their port facilities, and much of the surrounding area, levelled by an explosion. In video of the event, you can actually see the pressure wave, which is not often visible outside of nuclear explosions. Apartments three kilometres from the blast site had doors and furniture **inside** overturned and broken. Residences ten kilometres from the epicentre had windows broken. The sound of the blast was heard 200 kilometres away, in at least two other countries. The explosion registered on seismographs, at 3.3 on the Richter scale. The full story of the explosion is probably not yet known. Initially, the government reported that stored fireworks had exploded. Later reports noted that enormous quantities of ammonium nitrate were stored in the area, a cargo seized six years earlier. Ammonium nitrate is widely used as a fertilizer, but it is also the AN in the improvised explosive known as ANFO, which is simply Ammonium Nitrate (as the oxidizing agent) mixed with Fuel Oil (as the combustible material). Since ammonium nitrate is the oxidizing agent, it can be mixed with any fuel or combustible material, such as oil, flour, or simply any dust that might accumulate over six years, in order to form an explosive. Apparently there are also reports that welding was going on in the area. As we know from physical security (Chapter 5), for a fire you need oxygen or an oxidizing agent, fuel or combustible material, and a source of heat or ignition. Risk management (Chapter 1) tells us that when you have fuel and an oxidizing agent, adding a heat source without adding layers of preventive controls (Chapter 3) is just begging for trouble.

Be prepared. (You have to do that in advance, not when a disaster actually hits.)

The most important lessons are Dr. Bonnie's rules. Be kind. Be calm. Be safe. Helping others will help **you** more than filling your garage with toilet paper and locking it down.

We will get through this. We (mostly) help each other; working together, we can accomplish great things. And if cooperating with each other doesn't do it, remember that we are the most destructive life-form on the planet and have eradicated huge numbers of other species without even paying attention, so one virus doesn't stand much of a chance against us.

I am not a doctor. I am not a virologist. This is not a medical book. But I put myself through university by working in the medical field, first as a practical nurse (I spent considerable time working in an isolation ward) and later as an industrial first-aid attendant. I've also been an emergency management volunteer for a couple of decades. But this is not medical advice. I am a security professional, not a medical professional (any more). I will express opinions on some of the medical issues. I'd like to think my opinions on medical issues are informed, but they are only opinions.

One more medical point. 'Coronavirus' is a class or family of viruses that have similar basic physical structures. SARS-CoV-2 is one member of the coronavirus class and, before it had an official name, was referred to simply as a new, or novel, coronavirus. SARS-CoV-2 causes a disease called CoVID-19. A lot of people get these confused and think coronavirus, SARS-CoV-2, and CoVID-19 are all the same thing. They are related but there are distinctions. I may, in this book, sometimes use them interchangeably, and, if I do, I may sometimes use them the wrong way. If so, please forgive me. This book isn't medical advice, it's just using coronavirus, SARS-CoV-2, and CoVID-19 and the crisis arising from them as an illustration of security principles.

This book is based on a number of individual observations, lessons, short articles, and other bits and pieces that I wrote and noted as the crisis went on. You can write a book that's just a random collection of idle thoughts. (I've reviewed a number of them.) But it's tedious and difficult for the reader, and the lack of structure and discipline makes it more or less inevitable that the author will forget or neglect something important. I have therefore opened with a chapter using the CIA (confidentiality, integrity, and availability) triad and then I'll use the old ten-domain model that the International Information Systems Security Certification Consortium (or (ISC)[2]) used to use to arrange and organize exam and educational materials by subject, to structure most of the rest of it. (I know the newer eight-domain model is more holistic and therefore more representative of the reality of information security, but the ten-domain model, though arbitrary, is somewhat cleaner academically and easier to use for a book or course structure.) Hopefully, that will give us a place for everything and an easier way to find references.

Who is the audience for this book? Well, as I mentioned, it started out as lessons, notes, and reminders for my colleagues in the profession of information security. However, I'm a teacher, and I'm always thinking of my classes in security whenever I am writing anything. So, while this is not a complete guide to security, it is for students of security a kind of giant case study, pointing out many of the most foundational concepts of security as they have been illustrated by the current crisis. But, since it is a current crisis, this book is also a 'hot topic' and so there are probably some non-professionals and non-students in the general population who may be interested. And, in the general population, there is an ever-present need for security awareness. I've tried to write this in such a way that you don't need a formal security background in order to understand it and get something out of it.

Who is my competition? There are already some excellent security references on the market. The very best single volume is *Security Engineering: A Guide to Building Dependable Distributed Systems* (Wiley, 2001, 2008, 2020) by Ross Anderson. (Anything Anderson writes is worth reading.) Mine is not meant to compete with it. But, hopefully, a few people will be interested in the topic and maybe my book will be a little more fun to read (and you might learn some basics).

This is a new one on me. I've written books before but never one about a current event while the situation was going on. So, as I'm writing bits of it, new information is coming in and other pieces are changing. I hope I catch all the rapidly changing data before this book goes out, but with the rush, you never know.

I'd like to thank a number of people. Thank you to Dr. Alfred Kinsey for asserting that any field of study can yield value if only you gather enough data. Thank you to Dr. Bonnie Henry for constantly reliable and solid information on the state of the pandemic and the measures to take, as well as an example of supreme patience in the face of repeated demonstrations of ignorance. Thank you to Dr. Martin Wehlou (author of the excellent 'Rethinking the Electronic Healthcare Record') for noting my more egregious errors in medical terminology and concepts, as well as encouragement when I very badly needed it.

As a published author, I get various requests for advice from people who think they might like to write a book. One thing I tell them is that, if you find a good copy editor, marry her. I didn't actually know that I wanted to be a writer when Gloria agreed to marry me, and I didn't, at that time, know just how good a copy editor she was. God has given me the gift of an amazing and wonderful woman, whose talents are definitely not limited to copy typing, copy editing, and structural editing, the more arduous task of developmental editing. My first book was dedicated to her, and it is absolutely no exaggeration to say that I would never have become a writer without her.

Expanding Security

1

The CIA Triad

First, we need a little bit of history. Going back, way back, originally there was 'computer security'. This was the physical securing of the computer. Computers, back in the dark ages before you were born, were huge and expensive machines. They needed power, they needed air conditioning, and they needed meddlesome people kept away from them since they were also fairly delicate but unusual in construction and operation.

Over time, computers became more common, cheaper, and also more robust. As computers grew cheaper and as the storage and memory systems became larger, even more data was stored on them. Eventually, people realized that the data had a value and that the value of the data often exceeded the cost of the physical hardware. At this point, we started talking about data security.

Initially, we were concerned only with confidentiality. Could we keep the data safe from prying eyes? During this time in history, computers were expensive, and while business was using them, they were still mostly the preserve of the government and the military. The government and the military have always been concerned with people (mostly enemies but often their own citizenry) **not** knowing certain things. So the main concerns of data security were keeping people out and locking the data up.

But, as more companies and people started using computers, many started to realize that there was no particular point in keeping the data locked up if it wasn't available to them. That, in fact, there was sometimes a very solid business value in making the data available to your customers and others. And, once they had realized the importance of availability, there was also a realization that even having the data available was no good if you couldn't rely on the integrity of the data: had the data been improperly modified in some way? Did the changing world mean that the data no longer corresponded to the real world? Could you be sure that the person the data stated had said or done something had, in fact, said or done it?

Confidentiality, integrity, and availability became the CIA triad: the three fundamental pillars of information security. For most professionals, infosec is not complete if it does not effectively address those three concepts.

(Some in the field—and I must admit that I'm one of them—feel that integrity is just a special case of availability, so all we really need is CA. On the other hand, a number of people think the CIA triad doesn't quite go far enough. Donn Parker, for example, has created what many refer to as the Parkerian hexad, which adds authenticity, possession, and utility to the original confidentiality, integrity, and availability. The hexad definitely has its value, but for the purposes of this book, I'm going to stick with the original triad for the structure of this chapter.)

Nowadays, there is a new term being bandied about: cybersecurity. Some feel that infosec, with its triad, is incomplete since it doesn't (overtly) address the extent and increasing importance of communications and networking in the modern world. Many of us dinosaurs feel that (a) the term 'cybersecurity' is a poorly defined marketing phrase, (b) the 'domains' of security—as defined by the International Information Systems Security Certification Consortium (or $(ISC)^2$)—have always had a networking component, and (c) cybersecurity doesn't (yet) have any structure to distinguish itself from infosec and so can't be used to create a structure for a book, chapter, or course.

CONFIDENTIALITY

A woman, dying of CoVID-19, repeatedly talked to 'Alexa' about it, particularly about the pain she was experiencing. This raises all kinds of issues in regard to 'digital assistants'. We have seen a number of instances where these various voice assistants have failed, sometimes drastically, in situations where emergency help has been required. (More generally, there is the issue of the quality of information provided. This latter issue might belong more properly to considerations of integrity.) There are also increasing instances of people 'socializing' with their devices (which, in these days of quarantine and isolation, must be a concern). (See also the movie *Her*.) The point that I am trying to make here is that, particularly in a disaster (like a pandemic) that isolates people from each other, many are turning to their devices in order to have some form of social contact, even if it is artificial. It is probable that many people are giving away to these devices (and, more importantly, the corporations behind them where all the information actually resides) much more information than they would have been comfortable with before the pandemic began.

There are also issues in regard to the ability of users to interact with the devices and understand the reality of their capabilities and responses. (My mother is in a care facility, which is, of course, keeping visitors away. I have a Facetime call, which the staff will be handling, booked with her in a few hours. My mom does NOT handle technology well.)

SECURITY, PRIVACY, AND CONTACT TRACING

Because security is not a single 'thing', its parts sometimes come into conflict with each other. Confidentiality and availability are often in conflict, almost inherently so. When you are responsible for security, you are doing a balancing act—often several such acts at once.

Well, has there ever been a more glaring conflict between security and privacy? We, globally, are in the midst of a pandemic. Contact tracing is one of the absolutely vital tools for containing the spread of the virus (particularly since many of the testing tools that we would otherwise rely upon are slow, error-prone, or in limited supply; we will talk more about testing and errors later). Public health and our populations as a whole need (rather desperately) information not only about specific community transmissions and infections but also about patterns, processes, and actual risks of transmission, and masses of contract tracing data would be vital to providing the basis for this research.

At the same time, certain governments (probably **all** governments if they thought they could get away with it) are absolutely salivating at the thought of the surveillance possibilities they could gain through contact-tracing data. Yes, Google and Apple, in their cooperative work on the issue, have just (as I write) strongly come out in opposition to location data being part of contact tracing, but location should not be too hard to infer from the masses of contact data that could be gathered. Certainly, your familial, social, work, and other contact data would allow all kinds of surveillance possibilities. Whom do you meet? Whom do you meet **frequently**? Whom do you meet **surreptitiously**? And, since de-anonymization is so easy, what troublesome ideas and philosophies do they have that you might share?

Governments aren't the only ones. Companies would love to know whether you are social or a homebody and just how social you are. Simply looking at the number of close contacts you have would give them lots of information about the types of things you do and like and might be tempted to buy. And do you have frequent close contact with one (or more) of their existing customers? Marketing gold.

At this point, I should note that there are some significant differences in approaching the general topic of contact tracing, depending upon whether you are considering the traditional function, which is undertaken primarily by people interviewing infected people and their contacts, or whether you are primarily discussing contact-tracing technologies, such as those that might be used in a smartphone application. Human contact tracers collect an awful lot of personal information. In fact, they may collect much more data than any contact-tracing app, and much of the information they obtain may be irrelevant to medical or epidemiological needs. By the nature of human contact tracing, the interviewer tends to ask a lot of questions of any contact to try to tease out details of locations and specifics of activities which the interviewee may not initially bring to mind. However, a human interviewer will likely have a sort of 'attentional blindness' to any information that is not actually relevant to the spread of the infection. A contact-tracing app, on the other hand, collects only very specific data but then holds that data forever unless specifically told to forget it.

I've structured this book partly on the old ten-domain model of security, and when I started to put it together, one of my first thoughts was that I could forget cryptography. Not so. Trying to build a protocol that provides for (effective) contact tracing but doesn't collect or give away too much private information about either the individuals carrying phones or the overall patterns of contacts that they have is, as they say, a non-trivial task. Any kind of system relying on a central server or database is asking for trouble (in terms of data mining, attack, and single point of failure). But creating a distributed system for effective and reliable contact tracing is a bit much to get your head around. Creating a digital 'cash' seems like child's play in comparison. (No, I'm not talking about the current crush of crypto-currencies with their blockchains and the like. These carry lots of contact and even personal data, even if it's mostly hidden. True digital cash attempts to be anonymous.)

Getting the correct balance between effective contact tracing and privacy is not important just for the privacy considerations. There is also the issue of buy-in. If you are already an authoritarian government, you can just order everybody to install the app and turn it on. (Of course, even then, you are faced with those troublesome people who manage to find loopholes in pretty much any computerized system.) But for those in supposed democracies, you have to at least pretend to give your citizens some choice in the matter. That could be a problem for the medical and health security side. If you can't convince a certain proportion of the population to participate in contact tracing, then the protective value collapses. (It's rather like security as a whole in that regard. Partial security very often tends to be no security at all.)

In researching this issue, I came across a couple of Websites that seemed to be trying to do contact tracing. Having assessed that they were somewhat legitimate and not just trying to feed malware, I checked them out. One seemed to detect that I was in Canada, but (I assume) you can select Canada, Mexico, or the USA. The first screen asked if you are feeling well or not. If well, it asks if you've received a flu shot, how old you are, gender, and partial postal or zip code. (When you drop though from that, it asks for your cell phone number to send you reminders, but that isn't mandatory. I assume that the same is true if unwell, but I didn't want to deliberately feed it false data.) If unwell, it asks you to specify from 13 symptoms (or 'other').

Are the Web-based systems useful? Good? A problem?

The use of most contact-tracing apps is (at least somewhat) voluntary, and I can't see any form of integrity checking, so I don't know how susceptible such systems would be to attack with false or flood false data. At the moment, I can't see evidence of any privacy issues. (I didn't give the one I tested my phone number.)

Of course, there are people who **will** want to deliberately fill whatever system is in place with false data. The type of people who write their names on bathroom walls will take a burner phone with a contact-tracing app, deliberately place it near the phone of an infected person, and then hide it behind a seat on public transit so as to falsely create as many contacts as they can. A sufficient number of such actions might create a kind of denial of service of the contact-tracing system by flooding it with false data to the point that real data gets lost in the fog. Nation-state actors might deliberately mount such an attack simply to generate confusion and mistrust in a certain jurisdictions' medical and medical management systems.

Contact tracing, so far, is being implemented by individual jurisdictions—sometimes not even nationwide but on a state or provincial level. Once people start travelling again, are any of the various contact-tracing apps or protocols going to talk to each other?

There are also technical issues of background operation, access requirements, function requirements, and power consumption which present major problems to be solved. Most of the apps seems to rely on broadcasting beacons or nonces via Bluetooth to be picked up by other phones in the vicinity. The thing is that broadcasting, of whatever type, is one of the most power-consuming functions on a phone and therefore drains the battery fairly quickly. Since this affects the overall operation of the phone, Bluetooth is often turned off by default, and the user who wants to use Bluetooth functions needs to turn it on. In fact, Apple and Google (as the major vendors of smartphone operating systems) have fairly strict limitations on permitting any applications to turn Bluetooth on. In addition, smartphones, unlike computer operating systems, do not deal easily with multitasking or multiple applications running at once.

Therefore, having a contact-tracing app, which needs to beacon (and listen) constantly, requires a major concession on the part of the operating system vendors.

A lot of factors affect the range of Bluetooth. Bluetooth is therefore not good at judging distance, and distance is one of the major data points that contact-tracing apps are going to be interested in. The Bluetooth protocol is intended to be used with devices that are within a couple of metres, but I've been able to get them to work from 50 feet away. In addition, Bluetooth can easily penetrate Plexiglas and other types of walls and barriers to viral transmission, so many false positives are going to be triggered by an infected person who is physically close but behind a shield.

Yet another fundamental issue in designing any technical contact-tracing system is whether to make it centralized or decentralized. A centralized system has definite advantages in terms of access to and searching of data. But it has three inherent problems: privacy issues, a single point of failure, and jurisdictional incompatibility. These can, of course, be addressed. In terms of privacy, you can put all your eggs in one basket if you truly trust someone to protect that basket. A single point of failure can be made resilient or duplicated. And different jurisdictions can agree to mutual cooperation (although all too often they don't). Decentralized systems have their problems as well. As noted (and as we'll discuss further in regard to cryptography), creating a system that reliably gives you the information you need, but protects the privacy of those involved, is not easy.

There are, of course, other aspects of privacy and confidentiality that have been impacted by this crisis. In the rush to address various emergencies, there are always errors. The North Bay Parry Sound District Health Unit made a configuration error on a Website that exposed data—including names, identification numbers, and infection status—for 3,000 residents who had been tested for CoVID-19. These things happen, particularly in an emergency. Sometimes availability trumps perfection.

INTEGRITY

Errors

Sticking with contact tracing, let's look at errors. There are many types of errors, but in security we tend to think of two in particular. There are false positives and false negatives.

A false-positive error means that the system, whatever it is supposed to do or find, reports that there is a problem when there actually isn't. In terms of an antivirus scanner (for computer viruses), this means that the scanner tells you that your computer is infected by a virus when there is no virus. In terms of firewalls, it means that your firewall tells you there has been an attack on your network when there actually wasn't. In terms of a contact-tracing application, it should tell us when you have been in the vicinity of an infected person. And by 'in the vicinity', we can use Dr. Bonnie Henry's phrase of 'location, duration, relation'. You should have been close enough to that person (within two metres or six feet), for long enough (say, ten minutes), and doing something with them that is likely to get you infected. (This last one can be difficult for an app to figure out.) A false positive for contact tracing might tell you that you have been close enough to get infected because you were at the store and the clerk was infected whereas, in fact, you were separated from the clerk by a full Plexiglas barrier and didn't exchange any physical objects and there is no way you could have been infected.

A false negative is, of course, the opposite. The system doesn't alert you to the fact that there was a contact that might have infected you.

In regard to the contact-tracing apps on smartphones, a number of issues can lead to false positives. The location data for the apps is usually based on GPS and Bluetooth, neither of which is particularly reliable when it comes to determining distances of from 1.8 to 2.2 metres between phones. Neither of those technologies will be aware that you are behind a wall, curtain, or partition. In addition, you can be close to someone, for some time, and still not get infected: risk always involves an element of probability.

There are a number of reasons why your contact-tracing app might not tell you that you had contact with an infected person. There is the aforementioned issue of the accuracy of the distance. (Plus the fact that the virus can travel farther than two metres: it's not likely but it is possible.) The app might not properly transmit or copy the random code that is exchanged with most apps. Then again, the infected person you are near may simply not have a tracing app on their phone or (as is very likely with everyone designing their own) not one that is compatible with yours.

Now, you might think that of the two problems, a false negative is the greater evil. After all, you want the app to tell you if you are at risk and it might not do so. That's bad. And you actually have no idea if you are safe or clear.

But, actually, in our current situation, false positives are the bigger problem. As I write this, relatively few people in the world have been infected and even fewer are still infected and infectious. By the time you read this, the same situation is likely to prevail: the numbers won't have changed that much. Even in Wuhan, China, less than 1% of the population has ever been infected. Elsewhere in the world, it is much less, and in terms of actually infectious people, even less than that. Whatever the false-positive error rate is for the contact-tracing app, it is bound to be much more than 1%. So, for any contact-tracing app, if it gives you an alert that you might have been in a situation where you might have been infected, the alert is much more likely to be a false positive than a true risk. Knowing that, are you going to go into quarantine for two weeks every time you get an alert? Even if the government mandates quarantine on that basis, we are going to be quarantining many, many more people who are not sick than those people who are. Lots of workers, who may be important to their companies and society, are going to be forced to stay home.

The same holds true for tests for the virus. At the moment, everybody is eagerly looking forward to serology tests for CoVID-19. These are tests (usually blood tests) that determine whether you have antibodies related to defence against the SARS-CoV-2 virus.

At least, they try to determine that. Because, well, errors.

If the test has 99% specificity and you live in an area where only 1% of the population is actually infected, then when you get a 'positive' test and are reassured that you are immune, you actually have only a 50/50 chance that you encountered the virus and do have any defence. (In British Columbia [BC], where I live, the infection rate is about 0.03%, so the chance that a positive test is of any use at all is far worse.)

And this initial discussion of errors leads nicely into a more general discussion of integrity as the next pillar of security.

There is no point in keeping your information safe from prying eyes, or theft, if it is incorrect. There is a reason that the most widely known mantra of the computer era is 'Garbage in, garbage out'.

So many problems related to computers and systems can be traced to problems with information being incorrect. Maybe it was wrong when it was entered. Maybe it was incorrectly modified. Maybe it was incomplete. Maybe it was false information that someone fed you, hoping you would believe it and act in error.

The CoVID-19 pandemic has created so many examples of this problem. We are faced with a new virus and our information about it is so very paltry. We don't even know much about the class of virus. What is a coronavirus?

(I thought it was called coronavirus because of the 'crowns' of protein spikes that stick out of the fat layer. Speaking of information integrity, I was wrong. Back when the scientist who discovered the class of coronavirus named it, we couldn't yet see that level of detail. What she saw, at the resolution available on the electron microscopes of the time, was that it had the appearance of the corona around the sun when the sun was viewed through a high cirrus cloud. Canadian content time: the name is because of the electron microscope, a Canadian invention.)

As I write this, Donald Trump has announced that he has been taking the drug hydroxychloroquine. Is this useful information? Not really. All it tells us is that hydroxychloroquine doesn't kill everyone who takes it, and since it is a medication used for a number of illnesses, we already knew that. Is the drug safe? A number of studies have shown that it definitely has risks and should be managed by a physician. Is it effective as a treatment? A number of studies have shown that it isn't. Is it effective as a prophylactic against the infection? A study on that is under way, and we have no real data. (By the time you read this, maybe we will have.) It's been two months since Trump started pushing the idea that hydroxychloroquine protects against CoVID-19, and all we've really learned in that time is that it has become harder for people who really need it (for other conditions) to get it. Plus the fact that the US has stockpiled 63 million doses of the drug and has no particular use for it.

The plural of 'anecdote' is not 'data'. You will have heard reports stating that most people will experience only mild symptoms if they get infected with SARS-CoV-2 and develop CoVID-19. You will also have heard news media reporting people who have had CoVID-19 telling us that 'It's the worst illness I've ever had!' The thing is, both those statements are absolutely true. Statistically, over a large population, most people who get infected will have either no symptoms or very mild symptoms. But some get very, very sick indeed and some die—mostly those who are elderly or have underlying health conditions. Then again, there was a lady in Spain, 113 years old, who survived. She's an anecdote. When you are planning what to do about an epidemic, rely on the statistics.

Speaking of Donald Trump, it has become almost a truism that we are living in a post-truth age. Particularly in terms of the virus, there's a lot of misinformation being bandied about. There is a lot of erroneous information being distributed. (Then there are simple errors. At one point, the famous Johns Hopkins 'dashboard' of CoVID-19 case and fatality data accidentally dropped Canada from the list. At the time, Canada didn't have many cases and probably nobody cared, but it was a bit disconcerting if you actually lived there.)

Facebook says it is trying to get rid of fake CoVID-19 reports and 'miracle cures'. It seems to be having problems and is marking some legitimate postings and information about the virus as spam. Research has found that communities on social media that distrust establishment health guidance are more effective at influencing 'undecided' people than all the official medical or governmental agencies.

As it has been truly said, in many different ways, a lie can run around the world while the truth is still getting its pants on.

COVID-19, IBUPROFEN, AND INFORMATION INTEGRITY

It can be hard to determine what is legitimate information and what isn't. And even legitimate information can be misused or misunderstood.

Some of you may have heard that you shouldn't take ibuprofen since it (a) makes CoVID-19 worse somehow, (b) makes you more susceptible to CoVID-19, or (c) interferes with Ayurvedic cures. (OK, I made that last one up. But the way things are going with coronavirus misinformation, I expect to hear something like it any day now.) (By the way, if you think that there are any Ayurvedic coronavirus cures, you can stop reading now since you are obviously too far gone.)

The idea of a problem with ibuprofen seems to have hit the media through some public statement from a French politician. But it isn't exactly fake news, just an unverified hypothesis that so far has extremely little actual evidence behind it.

It's based on a 2020 article in the *Lancet* by Lei Fang, George Karakiulakis, and Michael Roth (https://www.thelancet.com/journals/lanres/article/PIIS2213-2600(20)30116-8/fulltext). The gist of it is (and please bear in mind that while I took biology and human physiology in university, I am definitely not a microbiologist, nor do I even play one on TV) that there is an enzyme called angiotensin-converting enzyme 2 (ACE2). It is produced by the outside cells of the lung, intestine, kidney, and blood vessels. SARS family coronaviruses (of which SARS-CoV-2 is one) bind to cells using this enzyme. ACE2 production is increased in patients with diabetes (who seem to be more susceptible to CoVID-19), and CoVID-19 attacks the lungs. So it is reasonable to think that ACE2 is important to infection with CoVID-19. Ibuprofen also increases ACE2, so it might be reasonable to assume that ibuprofen might make it easier for you to get CoVID-19 or you might get it faster or you might get it worse.

'Might'. But, so far, there doesn't seem to be any direct evidence for it. Even the authors of the paper say 'We therefore hypothesize' and go on to say 'If this hypothesis were to be confirmed'. Nobody seems to have observed any problems.

So it's important not to go beyond the facts. Or to spread information beyond the facts. You might, out of an abundance of caution, want to avoid ibuprofen. I don't take much ibuprofen but that's because I know and have observed and have direct evidence that, for me, it has unpleasant side effects.

So don't spread rumours. Verify information. Yes, you want to help, but it isn't helpful to spread unverified information.

Now go wash your hands. (And check your sources.)

There is misinformation being distributed. There is also deliberate *dis*information. Sometimes, the untruths arise from simple ignorance. There is a lot of ignorance to go around, particularly about a virus we didn't even know existed last year. Over the next few years, a lot of people will be doing a lot of study and research to find out more about coronaviruses as a class and about this one in particular. If only we knew now what we will know in a few years, we would be in good shape. But we don't know that yet. We are fighting, sometimes literally, for our lives (and the lives of those we hold dear) against an enemy we can only dimly perceive, and we have very little information about how it lives, attacks, and can in turn be attacked.

We fear the virus because it is dangerous. But we also fear the virus because we know so little about it. And that fear, arising out of our ignorance, can have a very strange effect on our decision-making. I saw this when I was studying computer viruses. In the early days, very few of us had a good understanding of how computer viruses worked, and most computer users and managers developed some very strange attitudes. Some, not knowing what viruses were or how to protect against them, decided (on the basis of no evidence at all) that the risk wasn't very big and viruses could be ignored. Mac users, particularly, developed a myth that Macs were immune and therefore they, as Mac users, were safe and need not take any precautions. (There were, in the early days, fewer strains of malware for Macs, but when they did hit a group of Mac users, the infection rates were much higher than in comparable groups of DOS users.) Many managers decided that viruses were a risk only if you used pirated software and, if you didn't, you were safe. (In the early days, the most prevalent and successful types of computer virus were the class of boot sector infectors, which travelled on floppy disks regardless of whether any software was present.)

We know a little about the SARS-CoV-2 virus. We know it is spread by droplets. We know most of those droplets fall to the ground or onto surfaces. We know that the highest risk of getting the disease is being in close (less than two metres) proximity to an infected person over a period of time. We know that it is unlikely you will get the disease from just one virus landing on you; you probably need a number of them (roughly a thousand?) getting onto or into your mucus membranes before you get infected. We know that the droplets produced by normal breathing put out about 20 virus particles per breath and that vociferous argument or heavy exercise (or, apparently, singing) puts out about 200.

Even in that limited subset of knowledge, there are so many things we don't yet know. We do know that some of the droplets are small enough that they hang in the air for some time, like aerosols. We don't know how many of those small, fine droplets have viable viruses in them. The tests that are available to us at the moment don't give us a lot of information about whether we are detecting a whole virus or just the parts of it that we are using as signatures. We know that masks, even N95 masks, aren't perfect, and we know that masks are better at preventing you (if you are already infected) from spreading the virus to those around you than they are at protecting you from getting the virus. We know that if you are in a particularly high-risk environment, wearing a mask is better than nothing. But we don't know how **much** better than nothing, and we don't yet have really good data about which high-risk environments are those that a mask might provide protection in. Therefore, we are talking about risk (which we will discuss further later on) and, as we in security know all too well, risk always involves a level of chance and probability.

Because of the probabilities involved as well as the huge areas where we simply don't yet have enough information, even an expert is going to have trouble giving you a hard-and-fast, yes-or-no answer to any question. If you stay at least two metres away from everyone, are you safe? Well, not really. There are still surfaces that someone infected might have touched or breathed droplets onto. Or the droplets may be smaller or the air currents might be stronger. There is not a magical barrier at two metres, but there is a good chance that the distance will keep you safe. Or safer. Are you safer at two metres than one and a half? Yes. How much safer? Don't know. Is a two-metre distance a good rule? Yes. Is it perfect? No.

Vancouver has a rather infamous area known as the Downtown Eastside. In a confined area (30 blocks), there is a large population (roughly 15,000) of street people, most having underlying health conditions. When the pandemic hit, it was considered a kind of foregone conclusion that the virus would sweep the area and cause massive damage and a huge death rate.

Didn't happen.

Now a study is under way to try to find out why. It's possible that simply some kind of social isolation between the street population and the rest of society has kept them safe. But it may be due to some kind of immune system protection that could be further examined and developed. It'll be interesting to find out.

Because there are so many areas where our knowledge is limited and imperfect, experts may (quite legitimately) disagree on various points. Different ideas and the importance of certain ideas may come into contention. So, in regard to masks (which have come to be a **huge** source of contention and which we will return to in other parts of this book), one expert may (quite honestly and truthfully) say that masks do not provide much absolute protection in terms of keeping you from getting infected. (As Tevye would say, you're right.) Another expert, just as honestly, could point out that, in areas of high viral load, a mask (backed up by a gown and face shield) is better than nothing. (As Tevye would say, you're right.) Yet another expert might note that, in populations where the rates of infection are higher and because of asymptomatic transmission, wearing a mask demonstrates that you are concerned (to the point of personal inconvenience) about protecting those around you rather than yourself. (As

Tevye would say, you're right.) And then some random bystander would point out that all of these points of view seem to contradict each other. (To which Tevye would say, you know, you also are right.)

Because of the lack of knowledge, both on the part of the general public and also because the experts can't give 100% assurance (which is another thing that we know in security: 100% protection is impossible), the fear comes back into play. Mr. John Q. Public doesn't know. And the experts he is hearing on the news media don't seem to be able to give single, solid, consistent, one-size-fits-all answers. So his decision-making starts to fixate on concepts that are backed, if at all, by very flimsy evidence.

In evolutionary terms, this is actually a good thing. Over a large population, the people who fixate on good ideas tend to survive and so those ideas become part of either brain patterns or culture and get passed along. Those who fixate on, well, not-very-useful ideas tend to die and remove themselves from the gene and culture pools. If you've got hundreds of thousands of independent individuals and many years to do the weeding, it'll benefit the species.

Unfortunately, in a modern, complex, and interdependent society, with a disease that arises out of nowhere and spreads rapidly through the population, fixating on random concepts just isn't very effective. You don't have the time for evolutionary pressures to work, and random chance is still very much involved. Some people with good ideas get unlucky and die. Many people with really stupid ideas survive, simply because they haven't encountered the virus in a way that has all the factors necessary for infection. And because our society is so interdependent, you, having good ideas, may still be at risk because you have to rely, for many of the essentials of life, on people who may believe that there are Ayurvedic cures for the disease and therefore don't think that handwashing or physical distancing is important.

But errors, ignorance, fear, and even the honest contention of ideas don't cover all of the untruth that is being spread about CoVID-19. Unfortunately, as we have seen during every major disaster, there are those who are willing and eager to spread lies for their own benefit.

There are many such frauds. One type of fraud is simply clickbait. People create Websites or mailing lists or post on social media with an attractive (to some) or appalling story. They usually say that the story is true. The point is to get people to visit a Website. The fact that people visit the site drives numbers that are used, sometimes completely fraudulently, to claim advertising revenue. In other cases, the Website may be used to spread malware. Or sometimes they are setting up supposed charitable sites and simply raking in the donations. Or, under the guise of signing people up for a cause or campaign, to phish for personal information, which can then be used in further frauds.

A huge attack on Washington State and its unemployment system netted hundreds of millions of dollars for a gang (probably based in Nigeria). The gang used personal data gathered from previous data breaches, and possibly phishing attacks, to generate huge numbers of phony claims. The US Secret Service issued an alert, noting that North Carolina, Massachusetts, Rhode Island, Oklahoma, Wyoming, and Florida were also subject to attacks. Part of the scam, in this case, relied upon the fact that regular postal mail was used as part of the check on fraud in the system but, owing to the pandemic, postal mail was being delayed.

At about the same time, here in Canada, we found out about a completely evil scam where the fraudsters would call up seniors and, for a fee, help them apply for the Canada Emergency Response Benefit (CERB). The CERB is a program for laid-off workers and applying for it is free. Not only would the seniors be out the fee, but if they did manage to get the benefit, they would be on the hook to pay it back since they were ineligible.

Sometimes, the stories themselves are attacks. Many nation-states are known for disinformation. In the past, fooling your enemy was seen as a legitimate tactic of war. These days, many nations seem to see themselves as constantly at war—with everyone. Therefore, creating a story that keeps another government off balance is a good thing. Creating a story that keeps the other country's population off balance, and possibly distrustful of their own government, is possibly even better. (So, even if you don't like the party in power, don't be too quick to believe anything bad you hear about them. You may be being manipulated by someone else. The Carnegie Mellon study found that most of the postings from bot accounts were aimed at creating division in America.)

Sometimes, it isn't even to their own benefit. As the saying goes, some people just want to watch the world burn.

AVAILABILITY

COVID-19, Toilet Paper, Hoarding, and Emergency Preparedness

Toilet paper? Really?

Of course, I'd seen the news stories showing streams of shoppers with carts full of toilet paper. The news stories all showed Costco, so I was hoping that maybe it was only a Costco issue. But, no. On my way home one night, I stopped for some groceries and the toilet paper aisle in my local Save-On was pretty bare. And it got worse.

Hoarding is a particularly insidious threat. It's hard to protect against. Unless you're going to ration, how do you tell people what (and how much) they can and cannot buy? Yes, I know. Rationing smacks of socialism or some other type of non- or anti-capitalist system. But hoarding is the inherent weakness of capitalism: unrestricted, capitalism tends to concentrate capital, which then becomes useless. Now, we are faced not only with the coronavirus but with the CoVID-19 toilet paper meme virus. People see that there is a run on or shortage of toilet paper, so they run out and drive around (wasting gas) trying to buy toilet paper. Creating a shortage of toilet paper.

(It's particularly galling here in BC. We have trees. We make toilet paper. By the ton.)

Why toilet paper? I mean, I defer to no one in my admiration for the stuff. It is one of the marvels of the modern age. (Toilet paper and the Internet.) It has lots of uses besides that originally intended. But it has no magical medicinal properties.

Yes, I know. In the emergency management field, we have been trying for years to get people to build emergency prep kits. Enough supplies to tide you over for three days. Or seven days. Or, in this case, two weeks. Fine. I get it. But do you know how much toilet paper you use in two weeks? You don't need to clear out stores.

(I have noticed gaps in the canned beans section and also in the soup aisle. Although, for some reason, Campbell's Chunky soups are completely stocked. Personally, I like chunky soups.)

And, if you are going to build an emergency prep kit, during an emergency is not the time to do it. You have to put some thought into it. How much toilet paper do you use in a week? How much soup do you eat in a week? Do you eat soup? Yes, I advise you to build an emergency prep kit. But build one. Don't just rush out and buy toilet paper.

Besides, CoVID-19 is not going to be the 'stock up on water and canned beans' type of regional disaster. You will still be able to get Amazon to deliver toilet paper to you if you get sick and have absolutely no friends in all the world to take care of you. (They may want to drop it and run and you may have to keep watch on your Ring-camera-that-is-insecure-because-you-haven't-changed-the-default-password-have-you to prevent doorstep thieves from stealing your toilet paper, but they will deliver.) (So, by the way, will Save-On.) Travel is going to be a problem, and stocks of your preferred supplies may be a problem, and there may be lots of other problems. But toilet paper is not going to be a problem. Unless people hoard it.

My barber is quite fond of making outrageous and sensational statements, so when he told me about $80 bottles of hand sanitizer, I didn't believe him. Turns out he was (almost) right.

Somebody asked what would happen to all the surplus toilet paper once the crisis was over. Toilet paper, if kept properly, has an expiry date of 937,164 weeks and two days. Store in a cool, dark place. Toilet paper rolls will not breed if kept away from wild trees. Every few months, check for evidence of dust bunnies: if found, remove the dust bunnies with tweezers and store in separate cages according to nationality.

Security
Management

2

When I facilitated review seminars for (ISC)², most of the candidates would come in thinking that technical aspects of firewalls and such were the most important areas. Those of us who have been in the game for longer know that you can have all the technical tools you want, but if they aren't managed properly, you don't have any security.

CoVID-19 provides a direct and glaring example of this. In terms of medical tools, treatments, and the research and industrial capacity to make more, it would be hard to make the case that any country in the world has more than the United States of America. And yet, as I write this, the US has the highest numbers of cases in the world, the highest number of cases per capita in the world, and the highest number of CoVID-19 deaths in the world. (Hopefully, by the time you read this, that will not be the case.)

SECURITY THEATRE

Unless you've been living under a rock for the past few months, you've seen videos, on the news, showing workers (and sometimes huge trucks) in China, Korea, and all over the world, spraying (and often fogging) huge outdoor areas. Supposedly, this is doing something about the new (that's what 'novel' means) SARS-CoV-2/CoVID-19 coronavirus.

Now, washing your hands, and possibly surfaces, regularly is a good preventative measure. But this isn't washing. It's just spraying (or fogging). Spraying surfaces with bleach, if there is a risk that they've been contaminated or even just on a regular basis, probably will kill most bacteria and even most viruses. But this obviously isn't bleach since it's often being sprayed around food markets, and you don't want bleach all over your piles of fresh fish fillets.

I can't (and my opinion was seconded by an infectious disease expert) think of anything that you can randomly spray around large areas that would

(a) kill viruses and at the same time (b) not kill people (or at least seriously compromise respiratory systems).

Disinfecting sprays seem to have become an enormous business. As I write this, schools are reopening and the news is showing school desks and classrooms being sprayed. Airlines are trying to convince passengers that airplanes are safe and so there are videos of the insides of planes being sprayed (sometimes with 'electrostatic' sprays). A French company is selling booths that spray you as you walk through them.

What is the content of the spray? Nobody seems to be asking that question. As previously noted, when people are scared and perfect information is not available, we, as human beings, tend to grasp at straws. When people are scared, you, as a government, must be seen to be doing something even if the 'something' does nothing to address the actual risk. Mexico had booths like those the French company is selling. They sprayed a diluted bleach solution. The bleach was concentrated enough to discolour fabrics and irritate skin but not concentrated enough to kill many viruses. A colleague in Saudi Arabia noted similar booths that sprayed a dilute acetic acid solution that was not strong enough to kill the virus but definitely left you smelling like a heavy batch of fish and chips.

A new market has opened up for devices that shine ultraviolet (UV) light. UV does kill viruses and even bacteria and sometimes very effectively. Also, UV, in small doses, is safe: it's present in sunlight. A surface placed in direct sunlight will be disinfected of SARS-CoV-2 within about four hours. (A recent study found that 90% of the virus was gone, in summer, within half an hour.) The thing is, the part of the UV spectrum that does the cleansing (known as UV-C) is mostly filtered by the atmosphere. (Also by glass, as is used in windows.) An ordinary decorative 'blacklight' is not going to disinfect anything. And any UV-C source that is strong enough to kill viruses is potentially strong enough to do some serious damage to you.

When I worked at the hospital, we had a UV-C device for debriding wounds. I never used it. The only people who were allowed to use it were a specialized office in the physio department since, if misused, it could be very dangerous. A careless exposure of less than a minute could cause damage that would strip the skin off your arms. When I worked in industrial first aid, we were taught about 'arc flash conjunctivitis'. When arc welders are careless with welding goggles, the UV light produced by their welding can damage the surface of their eyes.

We are seeing many different types of UV devices being sold to disinfect different things. We are being shown videos of devices (even one walk-through booth) that shine a vaguely purplish light that is supposed to kill viruses. If you can 'see' the UV light that is strong enough to kill viruses, it is strong enough to damage your eyes.

SOCIAL ENGINEERING

Security theatre is doing—or, really, being seen to do—something that actually isn't going to have any particular effect on the risk or threat people are worried about. That doesn't mean it doesn't do anything. It keeps people calm. In some situations, that can be quite important.

In security, when we talk about social engineering, we tend to talk about attackers using it as a tool. We are simply using a fancy term for lying. Just about every form of fraud uses some kind of social engineering. Malware uses a version of social engineering via some kind of implied lie about what the program does (or doesn't do). Some of the biggest names on the black side of technology weren't really that technically adept: they were just good at lying.

But we can use the power of social engineering for good as well as evil. I'm a teacher at heart. I am always using social engineering, mostly to get people to try things they never thought they could do. If you are a manager, one of the two things you are always managing is people. It's handy to know some social engineering in order to manage them more effectively. (Someone has noted that people will run downstairs faster if you announce that there is leftover food from a meeting than if you yell 'Fire!')

Dr. Bonnie Henry, in her daily press briefings, was constantly being asked what fines or punishments there would be for infractions of, for example, directives about gathering in public spaces or non-essential travel to common tourist destinations. (We have a lot of those in British Columbia.) Her answer was always the same. She refused to name punishments and instead stated she was sure that most people in the province were trying to do the right things and to keep themselves and their loved ones safe. Dr. Bonnie employs social engineering. What she said is absolutely correct: most people do try to cooperate and follow policies and guidelines. But she also knows that you catch more flies with honey than vinegar. If you assume that people will not cooperate and therefore you set up punishments and negative consequences, you make people feel badly about the whole situation. In that case, people are less apt to cooperate and will tend to spend more time and effort finding loopholes and ways around the rules rather than finding ways to live within them. If you say that you are sure people are doing the right thing, they tend to feel positive about the situation and are less inclined to amplify any minor grievances into a major problem.

Use social engineering and get it to work for you rather than against. As someone else has pointed out, if we had called it the 'stay-at-home challenge' instead of 'lockdown' and posted it on social media, the virus would have been eliminated by now.

SECURITY SELF-THEATRE?
(COVID-19 AND MASKS)

One thing that has bugged me ever since it entered the news cycle is the issue of face masks.OK, particularly in view of the hugely divisive, contentious, and frequently changed advice about masks, I have to make several important points. First, masks do provide you with some slight protection if you are in a situation (such as public transit) where physical distancing is not possible to enforce. Second, masks do provide some prevention of transmission if you are infected (whether you know you are infected or not). Third, if you are in an area of extreme public transmission of the virus, having any additional protection is good. Fourth, masks alone, in the absence of other protective actions, will not keep you from becoming infected.

Masks won't keep you from getting CoVID-19 or any other droplet-borne virus. (At least, they don't reduce your risk very much.) The paper face masks provide next to no protection in this regard, and the N95 masks aren't much better. Droplet-borne viruses will still get on your skin, on your face, and into your eyes, and simple daily activities make you touch your skin, face, mouth, and eyes and provide the viruses a path inside. You don't need to inhale the virus to get it, and if you do get CoVID-19, it's as likely that it will be from some pathway other than inhaling it. This is why frequent (very frequent) hand-washing is important. (Hand sanitizer is good, too. If you use it frequently.)

This advice, by the way, applies to influenza as well. Which brings up another point: if you are worried about the CoVID-19 virus and haven't gotten a flu shot, you are engaging in risky behaviour. Even in China, you are much, much more likely to get the flu than CoVID-19. Even in China, the likelihood that the next person you meet will have CoVID-19 is about 0.0001. (Probably somewhat less.) But if you go out into a crowd (if you can find a crowd in China these days), you are likely to encounter somebody with the flu. Having a flu shot probably doesn't reduce your risk of getting CoVID-19, but it does reduce your risk of getting the flu. If you get the flu, then you may have to get tested for CoVID-19, and the fact puts that much more demand on the health system and resources.

Given the furore over masks, you probably **really** don't care what my opinion on masks is. What you should care about, though, is what it says about risk assessment and risk management.

The trouble is that, at the moment (and in the midst of a crisis), a lot of people—some authoritative but with specialized agendas, some not authoritative, and some merely visible and persuasive—are saying (on the basis of very limited evidence) that masks might be good for you and, besides, what could it hurt?

Let's look at the (remarkably few) benefits and the (much greater) risks. There are quite a number of reasons that it might hurt a lot.

The first is, what do you mean by 'masks'? There are dust masks, which are intended to keep you from breathing in relatively large particles like sawdust. There are surgical masks, which look almost identical to dust masks but are made differently and of different materials and to different standards, intended to keep you from breathing (or, more realistically, spitting) out droplets of who-knows-what over patients with open wounds. There are slightly form-fitting masks made of specially porous materials that provide a larger surface to breathe through and so filter smaller particles, droplets, and some aerosols out of the air you are breathing in. (These also tend to catch most droplets that you are breathing out, but probably not all, and not aerosols, since, when you breathe out, your breath tends to push the mask away from your face and allow your breath, aerosols, and some droplets to escape above, below, and to the sides of the mask. The same happens with dust and surgical masks.) Then there are extended form-fitting respirators, many with integrated face or eye shields and with filters to deal with specific particle sizes.

(I should also mention the various masks with valves. These masks have one-way valves to reduce the extra effort needed to breathe through masks, even though masks have not been demonstrated to actually cause hypoxia or other medical difficulties. In regard to the current situation, though, the major benefit is to block the droplets that you, or any infected person, are breathing out and that may be spreading viruses to other people. The problem is, those one-way valves are open to breathing out, so they are not as effective in the way that we want masks to be effective in the pandemic we are facing.)

And then there are homemade masks, fashioned from whatever fabric is to hand, to whatever design comes to mind, with little or no regard for porosity, thread count, or the ability to trap whatever particles are being breathed in or out. Some are cotton, which is cheap and probably relatively effective at stopping droplets that you breathe out. Some are artificial fabrics, threads of plastic, and probably less effective. Then there is the question of stitching folds or layers into the fabric of the mask. (Which is a nice illustration of layered defence: if your protection is imperfect, back it up with more, even if also imperfect.) I carry around a bandana just in case someone is upset by the fact that I don't wear a mask. When I need it, I fold it in half, diagonally, and tie it around the back of my neck, like a bank or train robber in an old-fashioned western movie. It covers from the bridge of my nose down below my chin and can even be tucked into the neck of my shirt for more complete coverage. Otherwise it sits in my back pocket.

I am not advocating for any particular style of homemade mask. Nor am I going to fault any specific pattern or design as such. But the major factor involved in using a homemade mask is that they aren't made to any specific pattern. Or with any standard of fabric. If a mask covers your mouth and nose, it will somewhat impede and possibly trap droplets you breathe out. But, beyond that, there isn't much you can say. How much such a mask reduces the risk of your transmitting the virus is open to question. How much such a mask protects you is even more so. (The one advantage homemade masks do have is that you can make a lot of them and therefore you can carry enough that they are 'single-use' and thrown into the washing machine after every use.)

On the nightly news, one news anchor proudly showed off a face mask that his wife had crafted. It had a lovely pattern on the fabric and was lined with plastic from a bag. Excuse me? Plastic? Non-breathable, non-porous plastic? I'm not sure what that is supposed to do. Any breathing is going to take place around the edges of the mask. A normal person, under no effort or stress, is probably not going to be harmed by it, but anybody with respiratory problems who uses it may be in serious difficulty. It may, almost accidentally, trap droplets that are breathed out, but otherwise I can't see any possible benefit at all.

So, when you just say 'masks', it doesn't give you a lot of information about the level of risk and the level to which the masks reduce the risk. The first lesson about risk management is that you can't say much about the risks if your situation and many of the variables in it are so very … variable.

But, I hear you cry, while all of this calls into question the effectiveness of masks, it still doesn't show that masks (other than the plastic-lined ones) are harmful. So who's hurt if I choose to wear a mask?

Well, first off, we currently have a worldwide shortage of proper masks (and other medical equipment). If you are wearing a mask and don't need it, you may be (and likely are) depriving some front-line worker who may actually need it. In fact, if you have a proper mask these days, you probably got it on the black market and you are, even if only in a small way, supporting criminal activity that extends up to massive theft of hospital supplies and the fraudulent production of 'certified' medical equipment that is not up to anybody's standard. So you are probably hurting those doing the most to keep us safe. (And from there on down to legitimate manufacturers and the legitimate economy.)

And, even if you have made your own (probably ineffective) mask, you may be hurting yourself. We know that frequent, even obsessive, handwashing is effective. We know that physical distancing and self-isolation are effective. We know that avoiding touching your face is important. Wearing a mask gives you a feeling of security and safety. An almost certainly false sense of security and safety. And if wearing a mask makes you feel more comfortable and you stop, or even reduce, constant handwashing or are less careful about physical distancing or go out more frequently, you are putting yourself (and likely others) at greater risk. And we also know that properly donning and doffing a face mask are non-trivial tasks, and most people don't know how to do these things properly.

Sometimes, the rules about masks are security theatre. In a number of jurisdictions, masks are now mandated when you're out in public. There still isn't any evidence that this makes a really significant difference in terms of preventing the spread of the virus. It does provide some protection, of course, but we don't really know how much. The most effective orders for preventing the spread of the disease would be to enforce isolation and distancing and to close bars, nightclubs, and casinos. But these types of orders would be politically unpopular. So, instead, political leaders can order people to wear masks, knowing that the orders may or may not be followed and may or may not protect the population, but at least the political leaders can be seen to be doing something (whether or not it works) and don't have the political downside of ordering the things that actually do work.

A further lesson of the mask situation is this: we, as human beings, are really terrible at assessing risk. This whole novel coronavirus/SARS-CoV-2/CoVID-19/pandemic thing is just too huge and complex for us to take in. There are too many parts to it. Medical parts. Public health parts. Social parts. Psychological parts. Financial parts. Governmental parts. All of them interacting in complicated ways. We have difficulty figuring out what's important or true and what's not.

But masks—masks are simple. We can understand masks. Or at least we think we can. So we fixate on them.

Partly this inability to properly manage risks has to do with our evolution. Our attention to threats was formed eons ago, when threats, to be important, had to be nearby, moving (usually charging at us), and fairly straightforward. Our world has changed since then. (Mostly, we changed it.) We now have to pay attention to threats that are physically far away from us, possibly moving very slowly (so we have to look far into the future), and often extremely complex. We aren't built for that kind of risk assessment. At least, not automatically. So we have to build and use risk analysis frameworks and plans in order to ensure that we do, accurately, figure out what is a danger to us. And, as the saying goes, learn from the mistakes of others: you'll never live long enough to make them all yourself. When someone sets out some rules, it may be that there is a very good reason for them.

- Wash your hands.
- If you didn't got a flu shot, get one.

- Don't panic-buy, horde, or misuse masks and gloves.
- Keep (six feet) away from me.
- Now go wash your hands.

'ACTUAL FACE' MASKS

Danielle Baskin had an idea for a silly art project. The idea has taken off.

Her concept is to print the images of partial faces on face masks. It could be your face. It could be a random face. It could be an artificially generated face.

There are many uses for the idea. It could be used to provide masks for medical staff so they aren't as scary to kids they are treating. Or they could be used to 'create' a face to unlock your phone, so that your actual image isn't being used for that purpose. (It could be used to avoid all kinds of surveillance in that regard. Lots and lots of people who live under authoritarian regimes are very interested. Of course, it could also be used to avoid legitimate identification so you can engage in nefarious activities.)

STATISTICS AND PROTECTION

Remdesivir works against CoVID-19!

Sort of.

When it comes to trials of this kind, you have to look at the details, not just the headlines. This trial does appear to have good design with randomization and a control group with a placebo. That's good.

The results, as reported so far, are positive. That's good.

For those who took the remdesivir, recovery time was shorter. That's good. It's particularly good in situations where the medical system is being overwhelmed and getting people out of the hospital is a priority. But the recovery time was an average of 11 days (versus 15 days for the control group). That's not exactly earth-shaking. Also, we probably need to look at the definition of 'recovery' and particularly look at long-term effects like ongoing respiratory and neurological problems that have been reported in some 'recovered' patients.

For those who took the remdesivir, mortality was lower. That's good. But the mortality was still 8% for those on remdesivir versus 11.6% for those on placebo. Again, not a result that you want to rely on when people start thinking, 'Oh, there is a treatment, so I don't have to worry as much about getting infected!'

CRUISIN' FOR A BRUISIN'

The travel industry is taking a beating over CoVID-19. First it was the airlines who were the bad guys, for spreading it around the globe. Then it was the cruise ship industry, for being floating Petri dishes for growing new cases and becoming virus hot spots all on their own.

Frankly, I'm a bit surprised. With all the norovirus activity on cruise ships over the past decade, I would have thought they'd have had a better handle on how to deal with highly contagious viruses specifically. You know: building with easily cleanable surfaces, having highly filtered ventilation systems— that sort of thing. But no, apparently they just sailed along, blithely throwing infected passengers off ships and trying to avoid giving refunds. Until now.

Once again, yet another example of how most people refuse to prepare, even when they've been given mild warnings that something might happen.

Until it's too late …

The various ships that were quarantined, or refused berthing rights, provided ample fodder for the news media to find discontented passengers who were only too willing to complain about their situations.

These various interviews got to be pretty boring after a while. They tended to sound the 'Somebody (else) has to do something!' drum pretty consistently. And then when somebody did do something, were the interviewees grateful? No, something was always wrong. 'We had to pay for our flights home!' 'We had to quarantine ourselves once we got home!' 'The quarantine accommodations provided for us weren't in four-star hotels!'

I'm sorry but, as an emergency management worker, I have pretty much zero sympathy for these stories. Emergency management is for emergencies. It isn't going to be perfect; it's part of a vastly imperfect situation. We are trying to provide you with the basic necessities of life. You have food and shelter and will survive the next two weeks. If you feel that your situation is not optimal right now, you might want to factor that in the next time you want to do something 'exciting'.

RISK MANAGEMENT

So people have lost jobs, friends, family members, life as they know it, and lives.

And the authorities are telling people, begging people, to stay apart and stay inside and stay away from crowds.

So what does Krispy Kreme (in New South Wales, Australia) do to aid things in this situation? They have a donut giveaway to celebrate their 83rd birthday. It was a success.

If you think drawing big crowds, in this environment, is a success.

RISK FACTORS

As has been said, we are all in the same storm, but we are not all in the same boat. As I noted in the Introduction, I am writing from a position of privilege. Even though an outbreak occurred three blocks away from where I live, management of public health in my province has been highly effective and I am at very low risk. (I can safely type those words since, if I die, I won't finish this book and nobody will ever read this.)

We do not yet know all about SARS-CoV-2 or CoVID-19. We do have some information about various risk factors.

First, there are risks regarding becoming infected. One is location. As many epidemiologists have pointed out, the pandemic is not universally and evenly affecting everyone. It is best seen as a series of local epidemics. Some locations have been hit very severely. Some, even some not far away from hard-hit areas, get off lightly.

We don't know all of the factors involved in geographic risks. Some have posited that hot and dry locations have less risk of transmission. Looking at which countries have been hardest hit might seem to support this idea, but it's hard to say for certain.

There are some indications that children don't transmit the virus to one another as well as older populations do. (Which kind of flies in the face of the point of faith among teachers that kids are little germ factories, spreading diseases every September when school reopens.) There are some other data that, combined with this observation, might support the hypothesis that children are at lower risk of getting CoVID-19 than adults.

Infection rates do seem to show that, across many countries, more women get the disease than men. However, this might be due to the fact that women are more highly represented in nursing and other 'public facing' occupations (such as retail clerking), so possibly this relates more to occupation than to gender.

Once someone has the disease, the risk factors are somewhat clearer. Those who are older have a greater risk of showing severe symptoms, have a greater risk of requiring hospitalization, have a greater risk of complications, and a greater risk of death. And, quite clearly from the data, the older you are, the greater the risk in each of these areas. Young people generally have less severe symptoms, lower need for hospitalization, and fewer deaths. Those in their thirties have more, those in their fifties even more, and those in their eighties are at extreme risk if infected. As always when talking of risk, probabilities, not certainty, are the order of the day. There are children who have died and centenarians who have survived.

Once infected, men seem to have worse outcomes than women do. The data may be slightly skewed by the fact that men hide problems and symptoms and therefore may simply be sicker once they admit that they are sick. However, death rates tend to confirm that men are more seriously affected.

A number of underlying health factors seem to be related to worse outcomes. Those with diabetes and high blood pressure seem to generate more hospitalizations, complications, and deaths. The same holds true for obesity, although obesity is also strongly tied to conditions like diabetes and stressed cardiac systems, so possibly this is simply the same risk factor being seen from different sides.

CoVID-19 has been a terror for the elderly, particularly for those in care homes. Not only do the factors of age and underlying medical conditions come into play, but also the factors of close proximity, lack of 'personal space', common eating and recreational areas, and staff constantly moving between residents. As generally provided in Western societies, elder care is almost a 'perfect storm' of risk factors for this particular disease.

There is yet another area of risk. We are used to infections where, once you have been sick and recovered, you are cured. In most cases of CoVID-19, this is the pattern we see. But there are some who do eventually fight off the virus infection and start to recover but still have lingering aftereffects. Many have respiratory symptoms and difficulties that persist for months. There are reports of weakness and fatigue. The loss of a sense of smell and taste can linger for some time. There are also reports of delusions and dementia. We do not yet have enough information to know how frequently this may be the case, nor how long it can last. In some cases, the disabilities may be permanent.

There are many parallels between risk in disease and risk in security. We do not know all the factors. We do not have all the information. We work, as best we can, partly in the dark. We try to tease out, as best we can, those threats that we can see, and we protect imperfectly against those we can only theorize about.

METRICS

Security supports business. Security is business, and requires management. And, as the old mantra has it, you can't manage what you can't measure. In security, we tend to refer to the measurements we make of things and situations as metrics. We tend to give lip service to metrics, saying that they are important (like auditors), but in reality we often try to avoid them whenever we can. We often measure the wrong things. One of the common examples of incorrect metrics is that almost all of us will measure the number of computer viruses that a scanning platform catches. However, the more important number is how many viruses it fails to detect.

One of the most helpful books in the metrics literature is *PRAGMATIC Security Metrics: Applying Metametrics to Information Security* (Auerbach Publications, 2016) by W. Krag Brotby and Gary Hinson. As you might suspect, PRAGMATIC is an acronym, reminding us that metrics should be Predictive, Relevant, Actionable, Genuine, Meaningful, Accurate, Timely, Independent, and Cheap. Dr. Hinson has done further work on the idea since then, and I'm very much looking forward to his next book on the topic.

The CoVID-19 pandemic has lessons to teach us about metrics. In British Columbia we get regular briefings and updates from Dr. Bonnie Henry, the provincial health officer, and Adrian Dix, the Minister of Health. Every state-ment starts with a standard list of metrics, and it is instructive to see what the different metrics actually mean.

The first numbers in every statement are the number of new cases, and the total number since the pandemic began. While these metrics are obviously important, right away we realize that possibly more important than the number of new cases is the trend: is the number of new cases more or less than it was the previous day, and for some days past. In other words, is the situation getting better, or worse? A recent addition to the statements is the number of the new cases that have not been confirmed through actual testing, but are, in the words of the reports, 'epi-linked': people who have been found by contact tracing and epidemiological study to be infected.

One metric missing at this point is the total number of tests that have been done over the past day. In relation to the number of new cases, this gives us the positivity rate for the testing. The positivity rate can indicate a number of interesting things, although you have to be careful to examine the situation carefully. If the positivity rate is too low, it may mean that you are testing the wrong people, and wasting resources. However, if the positivity rate gets too high (and this generally means more than single digits) it likely means that you are not doing enough testing, and you are probably missing people who, especially with a disease like CoVID-19 which may generate a number of those infected who don't show any symptoms, can be out spreading the virus.

The total number of cases is interesting, but equally important, and the next set of numbers to be given, are the numbers of cases who have completely recovered, and, therefore, the total number of people who are still considered to be 'active' cases. This is followed by the number of people in hospital, and in intensive care, which provides some idea of the drain and demands placed on the healthcare system.

The next set of numbers in the briefings is the total numbers in the five different health regions in the province. This reminds us, in security, that total numbers are all very well, but that some parts of our enterprise may be unequally affected, and we need to keep track of that fact. In addition, recently a separate category has been added to this part of the statement, noting people who are not from the province, and arrived here already sick.

The next metric is the number of deaths that have occurred in the past day. At this point, I should note that my use of the word 'case', when I am actually talking about people, is simply following standard medical reporting terminology, and not an attempt to hide the fact that each case is a sick person, and each death is a death. As Adrian Dix points out, every day when the report has a death to list, every single death means loss and grief for friends, relatives, and society as a whole. It is important to remember the actual implications of the metrics.

The final metric that is included with every statement reports outbreaks in care facilities, hospitals, and in the general community. In security terms, this is alerting us to vulnerabilities in the system. Seniors, in care facilities, are not necessarily more likely to contract the disease, but the impact and outcome tends to be much more serious. Outbreaks in hospitals are, of course, obviously serious in a number of ways, including the greater risk to the health professionals who work there, and on whom we all rely. Outbreaks in the community are important relevant to the situation: a large house party means somebody (actually a large number of somebodies) has been foolish and we can't do anything about that, but contact tracing needs to follow up on it. An outbreak in a workplace may mean that health and safety needs to look at the environment and processes in order to issue orders to ensure that businesses and industries may continue to operate safely.

COST/BENEFIT

Many jurisdictions around the world have experienced some form or duration of lockdown. This 'bent the curve'. Isolation from basically everyone reduced the fact of transmission. It reduced the spread of the disease, reduced the suffering, and reduced the number of deaths.

It has come at a cost. Quite a number of costs, actually. Some jobs can be done from home, but many cannot. That has meant unemployment for some, financial losses for others, and an enormous cost to the economy as a whole.

The financial hits ripple throughout our society. Wage earners lose their jobs. They can't pay rent. Landlords lose out on income. Cities don't get paid property taxes. That means budgets have to be slashed and more workers lose their jobs. Landlords also have trouble paying mortgages. Banks face a minor dip in their profits and so get bailed out by the government. (Wait, what?)

Some of the costs are not financial. Isolation prevents the spread of the disease and so keeps us medically or physically safe. But social isolation also has psychological effects and those in turn can have a bearing on physical and medical conditions. (Finance can also come back into play here. Poverty has a known and well-documented negative effect on physical health.)

As I'm writing this, many jurisdictions have 'reopened' and more are considering it. Most of the reopening is being done on a gradual or graduated basis. There is much discussion of a 'new normal' (which undoubtedly differs from place to place).

The disease has come at a cost. The preventive measures have come at a cost but have benefited in terms of lives not lost as well as grief unsuffered. The reopening will have benefits but it will also have its own costs in terms of suffering and lives. The reopening will allow business to resume, businesses to make money, and the economy to improve.

In security, we are constantly assessing costs and benefits in very tangled and interconnected situations. Security, as a profession, exists to support businesses, so one might assume that we would universally be on the side of reopening. But 'life safety' is always the highest priority in any security calculation if you want to pass the CISSP (Certified Information Systems Security Professional) exam.

There are no easy answers.

Access Control

3

DEFENCE IN DEPTH/LAYERED DEFENCE

One of the foundational principles of access control is the realization that you might fail. In fact, that any kind of protection you set up will fail at some point. No defence is ever going to withstand every attack mounted against it.

Therefore, we rely on a fundamental concept referred to as defence in depth (or layered defence). The idea is that whatever safeguard you put in place has some kind of weakness. You want to set up a compensating control that will, well, compensate for that weakness. You want to know what and where your existing weaknesses are and choose other countermeasures and controls that will provide support and backstops for those holes. Since your compensating controls may (will) themselves have vulnerabilities, we build multiple layers of defence.

So it is with the response to CoVID-19. A first step, when a new disease appears, is to try to prevent geographic spread. So we may try to restrict travel. We will ask travellers if they have been to an affected area, and we may not admit them if they have been. But there are holes in that policy. People may lie about where they have been or may have been infected in a different location because of transmissions we may not know about. So we do not rely on screening questions only. We do symptom checks. We may check temperatures since we know that fever is one of the symptoms of CoVID-19.

But temperature and symptom checks have weaknesses as well since people can be asymptomatic or presymptomatic. So relying solely on travel restrictions is foolish. We must prepare another layer of defence since we have to assume that the travel limitations will not be a perfect protection. We have to prepare our medical system to be alert for signs of the disease in our population, and we need diagnostic tests to confirm whether people in our area have become infected. But the tests, as we know, have error rates and detect only once people have become infected or ill. So we need to build an additional layer of defence in terms of preparing our medical system to handle a possible surge in infections.

Part of the preparation of the medical system is ensuring that there are adequate supplies of personal protective equipment for nurses, doctors, and others on the front lines of the system, who must inherently deal directly with infected patients. Another part is obtaining equipment and drugs that are known to have some benefit for patients with severe symptoms, such as drugs and ventilators. Yet another is clearing non-emergency patients and procedures out of the system, as far as possible, to build margin for a surge of cases should infections get out of control.

And so it goes.

Once the virus has become established in our populations (pretty much inevitably), we have other safeguards, such as advice or guidance (or, in some cases, directives) about the need for physical isolation and limitation of contacts. There is the physical isolation, which reduces the possibility of transmission between people. But that can't be complete since we still need to go and buy groceries. So if we can't isolate from others, at least we can stay two metres away from them. But we still need to open doors and pick up items we want to buy. So there is constant handwashing to kill and remove the virus when we touch those surfaces we must when we go out to obtain supplies. But we don't have options to wash our hands all the time we are out, so hand sanitizer is a fallback to handwashing.

I love public libraries. I was very sad when the libraries closed. So I was delighted when our public library managed to institute a 'takeout' system. They have created a system that has some strange components but is overall quite safe and yet very accessible. The first thing has to do with returning materials. It's going to be impossible to effectively disinfect books, so they don't even try. The 'drop box' system is still in place, and they don't touch any of the returned items for three days. Simple and effective. It keeps the staff safe and by extension the patrons as well.

In terms of taking materials out, they use the online catalogue and the existing 'holds' system. You place a hold on an item, and the staff place it on the holds shelf under your name. Normally, you'd get a notification and then come in, take the item from the shelf, and check out the item. Now, though, when you get the notification, you apply for a reserved time to come and pick up your items on hold. This reservation system reduces the number of people who will be coming to the library at any given time and therefore reduces the possibility of a crowd. The afternoon before your reserved pickup time, the library staff check out your items, eliminating the need for you to actually come in to the library, and place them in a paper bag, which reduces problems with handling and transfer. At the appointed time, at the entrance of the library, you hold your identification up to a window, so the staff can see it and tell who you are. This eliminates the need for an actual conversation (with attendant breathing and droplets) between patrons and staff. A staff person obtains the bag with your holds, opens the door slightly, and places your bag on a table outside the door, reducing direct transfer. Contact is minimized, everybody is safe, and yet library items are accessible to all. I think they've done a marvellous job and created a safe and useful system by layering existing systems and simple additions.

Part of the control system is detection. You cannot protect if you can't reliably identify what you are protecting against. In the case of the virus, this means we need tests to identify whether someone is infected. In the case of many diseases, this detection simply takes the form of diagnosis on the basis of symptoms. CoVID-19, however, presents some problems with this approach. For one thing, the disease is deadlier than most of those we encounter. For the very elderly, death rates can reach or even exceed 10%. Another issue in regard to normal diagnosis is that many (possibly as many as 20% in general populations) have no symptoms at all and that possibly half may have mild symptoms that might be ignored. Hence, the fact that temperature checks can reduce the number of travellers carrying the virus but not completely eliminate them. Some will have the infection but are not yet sick enough to have a fever. Some are infected but will never get a fever.

In this situation, specific diagnostic tests are required. The first such test was based on the genetic code of the virus itself. As we'll note in more detail in Chapter 8 on application security, these RNA tests do not identify the entire genetic string, but a section or sections that act as a sort of signature. These are the tests that use nasal swabs up in the nasal-pharyngeal area (or in the back of the throat). These RNA tests identify not only that you have encountered the virus but that you are producing it yourself. You are therefore infected. As previously noted, these tests are not 100% error-free. Viruses mutate constantly. To a certain extent, with additional types of genetic testing, this allows us additional information for epidemiological study and allows us to track where a particular virus strain came from and where (and sometimes how) it travelled. But it also means that, at times, the signature no longer matches, even though the virus is creating disease in much the same way. Also, the detection process is a complicated one, requiring functions for reproduction of the genetic code, specific matching, and some means of notification or indication. A failure in any one of these areas will mean that an infection is not detected. In addition, if the genetic signature is not chosen carefully, other viruses, related but not causing this disease, may trigger a false alarm.

More recently, additional serology tests have been developed. These are usually blood tests and detect (or attempt to detect) antibodies that the body will produce in response to an infection by the virus but unfortunately do not tell us a great deal about how many antibodies have produced, how effective they are against the virus, or how long they may last. These tests, when they work, tell you that someone has had the virus at some time but may or may not have it currently. Presumably, someone who has been infected will have some level of immunity, but we do not yet know how much protection that provides. Many such tests have been produced, and the majority of them have been demonstrated to have huge error rates. In addition, recent studies (not yet fully verified) may indicate that, particularly with people who have had no (or very mild) symptoms, the levels of antibodies may fall very quickly, and some people may have almost no antibodies after only a month.

In security, identification is an important and often difficult task. It is tied to the concepts of authentication (verifying the identity), authorization (the rights or limits related to an identity), and accountability (who is responsible for an action). When a test for a virus tells us a virus is present, we need to validate that this identification is true. We then use the results of the test to decide what action to take. We also need to account for the origin, at least the immediate origin, of the infection in order to determine and restrict spread.

Security
Architecture

4

OK, I suppose the first point to address is, what the heck is architecture? Architecture is one level of abstraction above design.

'Oh, thanks', I hear you cry. 'That's really helpful'.

OK, let me put it to you in more concrete terms. What is architecture? An architect is the guy who makes the plans for constructing a building, right?

Let's suppose you want a house. What are the parts we want to put together. OK, yes, you over on the side, in the pink shirt. Walls? OK, that's a good thought, but **this** house is going to be built on a little island in the South Pacific. We don't want walls since (a) they are going to trap humidity in and (b) they are going to keep the cooling breezes out.

Right, then, walls are out, but what other parts do we want for this house? OK, I know you are all scared now that I've turned down your first idea, but you know that I can stand here all day, and this book isn't going to go on until I get an answer. OK, yes, you in the back. A roof? Excellent. Basically, any type of house we want to build will need a roof of some kind.

Look at that word. 'Need'. Architecture is, very roughly, the refinement of requirements. What do we need out of the system? What do we want it to do? What do we really, ultimately, want to get out of it? Don't get too hung up on details: those will come later in design and implementation.

BEND THE CURVE, NOT THE RULES

I don't know who came up with that slogan—I first heard it from Health Minister Adrian Dix—but it's a great one. It reminds you that there is a reason for the rules. There is a requirement behind the rules. We need the rules. Because we need to bend the curve.

I assume that everyone is familiar with the phrase 'bend the curve', but the idea behind it may never have been clear, or the phrase may have been so dulled by repetition that it has lost meaning. There is a curve to every epidemic. The infection is introduced into a population. It starts with low numbers, very likely a single individual. Given that everyone who is infected infects more than one other person, the graph of the number of people infected in each time period starts off small and slowly but then increases and then increases very rapidly. At some point, (a) a vaccine is developed, (b) the population develops immunity, or (c) the population starts to die off, so there are fewer people to infect, and then the curve slows and eventually reaches a maximum peak. After that it drops down and tails off. (Until the next wave. Historically, there has never been an epidemic without a second wave.)

If you can trace and isolate every infected individual, then you can stop the epidemic, usually before it becomes an epidemic. This has, fortunately, happened with the Ebola outbreaks so far. With most other infections, and specifically with CoVID-19, this has not been possible. However, with certain safeguards in place, you can ensure that the rise of the curve is not as rapid and that the peak is not as high. This does tend to stretch out the curve, so that, in the end, similar numbers of people may get infected. Why, then, take those countermeasures if the same number of people are going to be infected?

The point is to reduce that peak. In any community, there is a limit to available medical resources. Hospitals do not have enough beds for every person in their catchment area. (Likely, there is roughly one hospital bed for every 2000 population.) Therefore, if a lot of people get sick all at once, the medical system will be overwhelmed. If people who have severe symptoms can't obtain treatment effectively, more of them will die. If the number of people who get infected is spread out over time, then there is less probability that the system will be overwhelmed. Therefore, fewer people overall will die.

In fact, 'bending' the curve probably does mean that fewer people will get infected. That is what happened in Canada. A number of the provinces and territories had relatively few infections. But four had a significant number. British Columbia (BC) had some of the earliest infections and outbreaks. But it also had the earliest health directives closing, for example, bars and casinos, and restricting gatherings of more than 50 people. Even though the directives were less draconian than in some other provinces, the fact that they were implemented early meant that BC, relative to population, had only a third of the infections of Alberta (next door) and even lower rates when compared with Ontario and Quebec.

Even in Quebec, the health directives ensured that the medical system was never overrun. And the fact that the curve is flattened means that more time is available for other activities, such as research on treatments and a possible vaccine.

Time is often vitally important in security. Winn Schwartau has written an excellent book entitled *Time-Based Security* (Interpact Press, 1999). It is based on the premise, noted in the previous chapter on access control, that we must assume our protections will fail at some point. Schwartau therefore asserts that the design of any protections must delay the successful attack. You can't guarantee that your safeguard will prevent someone from getting in or getting at your data or messing with your system, so you need to make the safeguard stall the attacker long enough to give you time to detect the attack and respond to it.

This is the point of bending the curve. Follow the rules, stay home, keep two metres away from other people, wash your hands. It won't (and can't) guarantee that you won't get infected but it will delay the spread and allow us time to do contact tracing, ramp up testing, and prevent the overrun of the medical system—time to find treatments that work and develop and test a vaccine.

FUNCTIONAL AND ASSURANCE REQUIREMENTS

When we say that security architecture is about requirements, we should note that there are two different types of requirements. The first type, functional requirements, is undoubtedly the one we think about most. Functional requirements are what we want the system to do, or not do, to protect an asset. For example, if we are talking about security management, a functional requirement might be a security policy. In access control, it might be a requirement to have passwords to protect accounts. (Don't worry: in a minute, we'll argue about whether passwords are any good.) In physical security, it might be a fence or a wall to prevent unrestricted access.

Assurance requirements are quite different. An assurance requirement is some way of determining whether your functional requirement is actually protecting you—if it is doing what you actually think it is doing. For example, in regard to our security policies in security management, we have auditors who read our policies and consider whether they are truly adequate or whether they have gaps that need to be addressed. The policy is the functional requirement; the audit is the assurance requirement.

In regard to passwords, we might run password crackers against our systems to see if some of the passwords are weak and need to be strengthened. We might also have system auditing that checks whether our accounts are being misused or abused in some way that indicates they have been compromised. Auditors (more technically oriented this time) might point out that passwords are an inherently weak system anyway, and we might want to move to multifactor authentication. A wall or fence might have an alarm or closed circuit television system that will detect whether someone gets over or past it. Some assurance requirements are there to assess whether your control is properly functioning; others try to determine whether the control is actually doing any good.

As you can see from these examples, you can have different types of assurance requirements for a given functional requirement. Also, different controls for addressing a given functional requirement may have factors and ease for providing assurance requirements. In a restaurant, for example, there is a functional requirement for hygiene and sanitation. Commonly, the control for this is handwashing. But, in that case, the only assurance requirements available are having signs in the washroom reminding staff to wash their hands or doing lab cultures on swabs from the hands of the staff (which would cost too much and take too long). Some restaurants, particularly fast-food places, now have their staff wear disposable gloves when preparing food. In terms of the functional requirement of hygiene, washing hands and wearing gloves both fulfil the need. But gloves provide an assurance requirement that patrons can actually see for themselves. (Yes, OK, gloves aren't perfect, and if the staff wear them between preparing food and other tasks or wear them too long without changing them, then both functional and assurance requirements are void.)

In terms of the pandemic, and particularly in terms of the public perception of steps being taken, the functional and assurance requirements are vastly different. The functional requirements are extremely visible, immediate, and inconvenient. Stay home. Don't go to work. Don't go to shops, bars, restaurants, or movie theatres or to visit Granma. Stay two metres away from anyone and everyone. Wash your hands. All the time. If your hands get chapped, use hand cream. (That's a kind of layered defence, remember? Or possibly a compensating control.) The assurance requirements, on the other hand, are invisible, long term, subtle, and inconvenient. You can't see the virus without an electron microscope. You can't see it passing between people, six feet away or not. You can't see it on your hands or surfaces, and you can't tell whether washing your hands kills it or washes it off. You can't see if you or anyone else is infected. You can see some symptoms, like coughing or fever (if you have a temperature 'gun'), but you can't see if someone is infected and asymptomatic.

Tests to detect the virus require labs and supplies, time, and training. Contact tracing really tells you only whom to test. Infection data really tells you only how effective you were at preventing spread of the infection two weeks ago. Death rates are definitely an after-the-fact indication.

SIMPLE

We tend to think of security as complicated, a complex interaction of convoluted systems. In a lot of ways that is quite true. But some pieces and parts of security can be incredibly simple. So simple they almost sound silly.

We've talked about masks. We haven't talked much about face shields, which are as important and protective as masks and possibly even more so in some situations (although, of course, they are complementary and the best bet is to use both). You will have seen pictures of people wearing face shields with headbands and hinges for flipping the shield out of the way when you need to do something that the mask impedes. Engineering companies, car parts manufacturers, and plastics specialists have retooled to produce them for the emergency.

One small shop had a slightly different idea. Used to producing small items for the restaurant trade, the owner noticed that a great many restaurant employees wore billed (baseball) caps. He designed a very simple shield with three clips that would attach to a ball cap. Voilà! The Cap Shield. (Patent pending, I assume.)

I'm not advertising or recommending his product. The point I am trying to make is that you don't need complexity to make security work. You don't even want it. Complexity is the enemy of security. If it works, use it. Even if it seems silly.

Business Continuity and Disaster Recovery Planning

5

It is always difficult to push for an increase in any business budget, and security is harder than most. Security is seen as a cost without any possibility of revenue generation. Security and business continuity planning (BCP) are seen as insurance: maybe you need to have them, but you want to pay as little as possible. It takes forever, and an awful lot of work and convincing, to start any new project. It's difficult when any existing work demonstrates an increased need for more resources. And BCP is possibly one of the hardest areas for which to get funding.

As many BCP specialists know, the best argument for doing BCP is having the business across the street burn down.

Well, right now, an awful lot of businesses across an awful lot of streets are burning down. Every week during the month of May in 2020, a major corporation filed for bankruptcy protection. Not just little mom-and-pop operations, which were going out of business so fast that nobody was keeping track. Early on, somebody said that CoVID-19 made all our business continuity pandemic planning redundant. CoVID-19 has a number of lessons for pandemic planning, but it certainly doesn't make BCP redundant. Check your supply chains. Find (and eliminate) single points of failure. Build redundancy. Do succession planning. The necessity for all of these has been amply demonstrated during the pandemic. They are all part of a real business continuity plan.

As I keep telling people in business continuity and disaster recovery seminars, use anything and everything you can to promote the need for disaster and emergency planning. And nothing works better to convince people than an actual disaster.

CAPITAL RISK AND MARKET CHANGES

When I began more formal study and teaching of security, one of the primary areas missing in the security literature was in the field of risk management. Of the literature that was available, much seemed to make reference to risk management guidelines and frameworks from the financial, and particularly the banking, industry. This was of limited use in teaching risk management for information security since the financial industry emphasized one area and one area only: capital risk.

(Actually, when I dug further into the banking industry risk references, I did find some mention of the type of risk analysis we do in information security—policies, technical vulnerabilities, attack profiles, error management, attack trees, spanning trees, models, and frameworks—all lumped into one little corner of their risk management structure, called 'operational risk'.)

More security people should pay attention to capital risk. The basic question is, do you have enough money to survive some sudden calamity? Ultimately, a large number of the problems presented by the CoVID-19 crisis fall into this category. Do you have a place to self-isolate? Well, yes, until you get evicted for not having enough money to pay the rent. Do you have enough to eat? Well, yes, the groceries stores are still open, but do you have enough money to buy food? We can start with people who lost their jobs (and didn't bother to save money when they had a job) and move up the chain. Can you continue your business? Can you protect your employees? OK, perhaps your business isn't set up for it now, but can you throw some money into Plexiglas dividers and personal protective equipment? What then? As the business mantra has it, if you have a problem that can be solved by throwing money at it, you don't have a problem.

Is personal protective equipment getting expensive? Well, do you have enough money to outbid other businesses? (I do **not**, you will note, recommend trying to outbid hospitals.) If you actually do have to shut down, do you have enough money to stay afloat for however long it will take? Do you have enough money in reserve to build that 'online presence' that you have been putting off doing?

Do you have enough money in the bank to weather a change in the market? In an instant, the bottom fell out of the markets for entertainment of all types: theatres, events, concerts, and any kind of tourism. That is, for live attendance. At the same time, those who are providing entertainment online are doing land-office business. Certainly, the market for live events will return, and when it does, it will probably come back with a vengeance, as pent-up demand plays out in huge line-ups for cinemas and clubs. But that time won't come until we have a vaccine, and a vaccine is probably at least a year or two away. Some businesses have been able to rearrange what they provide (and how) in order to accommodate the alterations in the market, but it takes money to get new resources that might be needed and money to survive until you are ready to relaunch.

The change in the market due to CoVID-19 has been wide-ranging, hitting many sectors at once, as society as a whole was forced to change the way we live. But market changes happen all the time and they are not always gradual. Examination of your capital risk, and the stability of your market, should always be part of a business continuity plan.

EFFICIENCY VERSUS RESILIENCE

For three decades now, I have had a feeling that our constant (business) pursuit of efficiency was going to turn around and bite us at some point. (In the press of other events and research, I haven't been able to study it as thoroughly as I would have liked.) Well, the CoVID-19 pandemic has amply demonstrated that efficiency is bad for security.

Dr. Martin Wehlou, noted in the introduction tells me that Sweden has had an extreme drive for efficiency in medical care for the last decade or two. Just-in-time (JIT) ordering of everything, abolishing stocks of material, the lowest number of hospital beds per capita in Europe. The result was predictable. Finland is a case study of the opposite. (As of August 2020, Sweden, which has a population of ten million, reported 81,000 CoVID-19 cases, the highest of any of the Scandinavian countries. Finland, which has a population of five million, reported 7500 cases.)

Initially, and specifically, efficiency eliminates redundancy and therefore is at odds with BCP. (As we say in security, a redundant backup is not redundant when you need it.) Our pursuit of efficiency and our elimination of margins in pursuit of immediate profits have created extremely brittle systems and supply chains. It has taken a global crisis to point out the danger. Unfortunately, it has put us globally in a business situation facing massive debt, which will take at least a decade to climb out of. A great many businesses will not survive.

It is possible that the failure of so many enterprises will force business management and economics to re-evaluate our devotion to efficiency and unrestrained capitalism as the only guiding principle for business. One can hope, but I do rather fear the old adage which asserts 'history teaches us that history teaches us nothing'.

PERSONNEL

Another part of business continuity and disaster recovery that gets short shrift is personnel. There are all kinds of ways to recover your systems these days and many options for transport, communications, and alternate supply lines (as long as you aren't looking for toilet paper or masks), but you cannot replace your employees in any reasonable amount of time. And it's your people who have been affected by this crisis. (And you too. We'll look at training and leadership in a bit.)

How much of your work can be done from home? How much can be moved to be done from home?

What support do your employees need to work from home? It's not just laptops and tablets and cloud-based apps: working on a laptop at a stool at the kitchen table for hours and hours is quite different from doing the same thing with a proper keyboard on a desk of the right height and sitting in an ergonomically designed office chair. Checking customer information and transactions at a desk with a large monitor and a phone headset is not the same as juggling your cell phone while trying to bring things up on a tablet.

And then there are those who could possibly have worked from home but probably can't WFH when that means Work From Hospital. Or, heaven forfend, some of your vitally important people might die. Succession planning is another area that the pandemic crisis has pointed out. Single point of failure doesn't relate just to systems: make sure that you can lose people (hopefully you won't) without losing your productivity as well.

SUPPLY CHAINS

If there is one single lesson that the virus crisis has taught just about every business, it is that JIT supply chains can be extremely brittle. Every industry sector has been hit by supply chain issues. Starting from the wrong end, as it were, in terms of changing markets, the entire petroleum sector, which most people always saw as a licence to print money, suddenly had no buyers, completely full storage tanks, and a price crash. Theatres couldn't show movies, but DVD and online sales couldn't sell new movies since the studios weren't releasing the movies since the theatres weren't open. In the northern hemisphere, the virus hit as planting season was starting, but travel restrictions meant seasonal farm workers weren't available. Some of the food supply was hit in very odd ways: there was lots of food, but all the food that was normally sold to restaurants was not moving, while grocery stores were having a hard time keeping shelves stocked because of (a) panic-buying, (b) food-processing plants being hit by virus outbreaks, and (c) all the people who would normally have been eating in restaurants who were suddenly cooking at home. (The occasional shortages of certain types of food fuelled more panic-buying.)

The supply chain has become extremely important and, at the same time, extremely complex. JIT supply, which has been very popular in terms of cost savings, has left very little margin of time for reactions when some disaster strikes. The radically increased demand for face masks and other personal protective equipment for the medical field, in response to the crisis, has pointed out just how complex the situation is becoming. An American company is one of the major suppliers of N95 standard face masks. At one point, there was a short-lived threat, by American authorities, to commandeer the entire supply of those masks for the American market, which would have cut off significant supplies to the Canadian market. Fortunately, saner heads prevailed, and nobody had to find out how well the masks could be made without the specialized fibre that comes from a mill on Vancouver Island. In Canada.

Similarly complex supply chains apply in regard to the reagents used in testing for the virus, in terms of both the RNA tests and the serology tests. And for pretty much every technical marvel we are coming to rely upon in our increasingly technologically dependent world. Phones sold by American corporations are made in China but require chips fabricated in the US—with elements mined in China. And that's a very simple supply chain these days.

Tying this back to capital risk, there is a grave danger of putting all your eggs in one supply basket for cost savings. JIT and single-source supply chains do trim margins and result in financial savings. But if that supply basket is jeopardized by a minor glitch in a complex web that is completely outside your control, your business may fail.

TRAINING

Those of us in the security communities are always interested in disasters. We are forever dealing with crises, both large and small, assessing risks, planning and comparing mitigation strategies, and looking at the management of it all. So I recall that, when Hurricane Katrina struck, there were endless discussions of the latest details, the structures, the organization (and lack thereof) in the follow-up efforts. One person made a donation to a charity and challenged the group to match his gift. Upping the stakes, I challenged everyone to sign up for training in preparation for disasters.

Unfortunately for the point I'm trying to make, I am again speaking from a position of privilege. Canada has the best emergency structure in the world. British Columbia (BC) has the best emergency response management system in Canada. (No, I didn't volunteer at the Olympics. But for the year before it, I worked with a group that was planning for the fact that, with the big event in town, even a minor crisis was probably going to mean that we might have to provide emergency lodging for a few hundred people.) And the North Shore, where I live, has the best disaster-training regime in BC. (The group-lodging thing wasn't done by the Vancouver Olympic Committee: it was an effort by the Emergency Support Services volunteers from the North Shore, Vancouver, and Richmond.)

Emergency response, in a major disaster, is not simply a matter of having water, generators, blankets, and rescue dogs. It has to do with organization, coordination, management, and (in particular) trained people. Most of them are volunteers since nobody can afford to pay for a full-time staff of all those you need to have ready in an emergency.

That's where you come in.

Get trained.

There is some emergency measures organization that covers your area. Your local municipality probably has an office. And they probably need volunteers. And they provide training. For free. (You may get additional perks. I get my flu shots paid for every year since I'm an emergency worker.)

First of all, you'll probably learn skills that you need for you and your family. What do you require to survive the first 72 hours following a disaster? Do you know how much water, what type of food, and so on, you'll have to have in store in the event of a total failure of utilities and other factors we rely on?

Then there are the skills that will assist you to help other people. Sometimes, this might relate to first aid, structural assessment of buildings after an earthquake, and so on. However, there are many necessary skills that are not quite so dramatic. Most emergency response, believe it or not, has to do with paperwork. Who is safe? Who requires care? Do families need to be reunited? Documentation of all of this is a huge effort, which goes on long after the bottles of water and hot meals have been distributed.

Then there are management skills to coordinate all of the other skills. An awful lot of 'charity' gets wasted because some people get too much help and others don't get enough. Someone needs to oversee the efforts.

Training in all of this is available. And, in an emergency, having trained people is probably more important than having stockpiles of tents. Trained people can make or improvise shelter.

Maybe your municipality or county doesn't have a formal emergency structure. In that case, there are organizations covering the gap. In Canada, the Red Cross and Salvation Army are two of the groups that have been working on this for years and have specialists. In BC, we have courses provided by the Justice Institute in a number of areas. The provincial government has created a marvellous structure, ensuring consistent organizational layout for all sizes and types of disasters and all types of response. But we don't bother reinventing the wheel. In our formal training curriculum, a number of the courses are prepared, provided, and run by the groups that have been doing it for years and know it best. If your government doesn't have the courses available, find those who do. They are around.

(For those who have security related certifications, like the CISSP [Certified Information Systems Security Professional], ongoing professional education is a requirement. Constant complaints are that training is expensive and getting the credits costs too much. All kinds of training related to business continuity and disaster recovery are available and most of them are free.)

Get trained. Volunteer. You'll get a wealth of experience that will help you plan for all kinds of events, not just for major disasters but for the minor incidents that plague us and our companies every day. You'll be ready for the big stuff too. You'll be able to keep yourself and those near to you safe. You'll be able to make a difference to others, certainly reducing suffering and possibly saving lives. If and when something major happens, you will be a part of the infrastructure necessary for the response to be effective. You'll be part of the solution rather than part of the problem.

LEADERSHIP

In all of this there has to be leadership. There is a requirement for a champion to get BCP going in the first place. Someone has to do the herding of cats that is typical of the planning process itself. The team must be driven through the tough times when it seems nobody is on your side. There is a need to encourage and acknowledge the team on the rare occasions when somebody does notice that they've done something important. And, most of all, somebody has to be in charge during the actual disaster.

Right now, we have an amazingly comprehensive example of how not to lead during a crisis. All you have to do is look at what is happening there and not do that.

A leader has to listen. You can't know everything: nobody knows everything. You need people around you who are gathering information, on an ongoing basis, and presenting it to you. The information is going to be imperfect since the whole situation is one enormous imperfection. You can't demand that they bring you only solid information. You have to be able to deal with ambiguity and realize that the people you are relying upon are facing problems of their own. You have to know enough about the various aspects of the situation to take incomplete and possibly erroneous reports, figure out what is probably happening, and still make decisions.

You have to be willing to make mistakes. You have to accept that you will. As G. K. Chesterton said, if a job is worth doing, it is worth doing badly. In any disaster, the best is very definitely the enemy of the good. If you are not willing to make mistakes, you are going to do nothing, and doing nothing is going to cause damage and suffering.

You then have to be willing to admit that you have made mistakes if later data contradicts the imperfect or erroneous information you had to begin with. Anybody who points out that something is wrong is your friend, not your enemy. And you have to be willing to get on with the job without trying to assign blame. It's not about you. It's about the job, the crisis, and the people who are in trouble.

You have to do what you can. And not worry about what you can't do. As the Serenity Prayer says, you must have the courage to change what must change, the serenity to accept what can't, and the wisdom to know the difference.

You have to be consistent. Yes, I know that that statement comes into conflict with the fact that you have to change and adapt as the situation (or your knowledge of it) changes. So you have to be consistent about staying aware, about supporting those who support you, about the focus on the primary goal. You are a leader: you cannot check to see what is popular today. You have to choose the right goal and get everyone to work towards it.

You have to motivate those who are doing the job(s) for you. You have to reassure them. Even if you're not terribly happy with how someone is doing their job, if they are contributing something, remember that you catch more flies with honey than vinegar. (Also, in the middle of a crisis is not a great time to be making important decisions such as whether to fire someone.) You also have to keep them aware of the real problems they are facing. You can't support people simply by saying that rainbows and unicorns are around the corner.

And you have to take care of yourself. You have to make sure that you are not becoming too tired, too stressed, too depressed, and too distracted and therefore starting to make bad decisions. You have to pay attention to yourself and your level of coping.

STOP FAST, RESTART SLOW

There are a couple of words that tend to trip up students of business continuity: recover and restore. Recovery is what you do when your business has been interrupted, in order to get back up and running: hence disaster recovery. Restoration is what happens when, after a disaster has interrupted your business, you have recovered (possibly to an alternate location) and are operating but now want to restore to normal functioning (generally in your original premises). So recovery is what happens during the emergency and restoration is what happens when you are finishing up after an emergency.

There is an odd and somewhat counter-intuitive difference between recovery and restoration. In recovery, you recover the most important and vital business functions first. Remember, recovery is what happens immediately (or as soon as possible anyway) after the catastrophe. You want to recover the most crucial and critical business functions and systems first so that you can get back to business as quickly as possible. But, in restoration, you restore the least important systems first.

Remember, in regard to restoration, that you have previously recovered your important systems. They are running. You are operating, even if the situation may not be optimal. In restoration, you are restoring your systems in your original (or possibly rebuilt) location. You therefore want to take the least important systems and start them up. That gives you the opportunity to make sure that hardware, power, and various other infrastructure processes are operating as they should to support your business systems. It also gives you the chance, while business operations are being conducted and supported elsewhere, to ensure that the systems you are restoring are behaving properly, and doing what they are supposed to, before you have to completely rely on them with a full move back to the original site.

Something similar went on with the lockdown, and reopening, during the pandemic. The lockdown (or various lockdowns) was imposed quickly by fiat. Society stopped overnight. But the reopenings, in most jurisdictions, happened much more slowly. Phases of reopening were planned and agreed to. Calculations were made, models were used, and data was gathered. The reopening (at least for successful reopening) was done much more gradually than the sometimes brutal lockdown. We recovered the safety of society by imposing a lockdown, because we had to. We restored society, and the economy, much more slowly and carefully.

Physical Security

6

There is, in physical security, a concept known as 'situational awareness'. As the name suggests, it is a simple idea, basically just being aware of the situation around you. Those who are really into physical security, which is generally the military and those ex-military personnel who do 'personal security' details, emphasize the topic. The theory is that if you are truly aware of the situation around you, you will know about any dangers before they get to you.

In reality, of course, we are all somewhat aware of our immediate situation (except for those walking around with cell phones). At the same time, it is impossible to be completely aware of our surroundings. There are limits to our perception, and we don't always have the knowledge to infer implications from clues that we can perceive. (In a foreign country where I had just arrived, I remarked that the farmers in this locale seemed very prosperous, surprising my hosts. A friend studying agriculture had once described a particular brand of silo that was both expensive and highly productive. Since there were a number of them in the countryside, it indicated that the local farmers (a) had enough money to buy them and (b) thereby would have increased their revenue.)

But it is true that most of us do not pay enough attention to what is happening in our vicinity. We are used to living in a society where dangers do not lurk around every corner, and we expect that the government or the police or various other authorities will keep us safe. We tend to expect this even if we don't particularly like or trust the government or the police or various other authorities (which I find rather inconsistent) and fail to understand that we are part of the government. (If you want the government to do something, maybe you should take the initiative. But I digress.) We rail against food agencies for not allowing unpasteurized milk, but we expect the food agencies to ensure that anything we buy (including from illegal roadside vendors) is safe and uncontaminated. We object to financial regulations but expect that we will be kept safe from fraudulent investment schemes or failing banks. It is likely that we should be paying more attention to what goes on around us.

(The strongest illustration I ever saw of this was at Disneyland. Nothing was 'real', and everyone expected everything to be completely safe. I got the strangest feeling that if you dug a pit, lined it with sharp spikes, and put up some guide ropes, people would line up to jump in and be killed.)

The lack of situational awareness has been amply demonstrated by almost any trip to the store during the pandemic. Various things about the way you shop have changed. There are Plexiglas barriers you shouldn't reach around, there are one-way aisles, there are barricades to prevent you from jumping the physically distanced queue waiting for the cashier. In addition, most people are trying to ensure that they do stay two metres apart from each other. Yet there are lots of people who blithely pass the hand sanitizer with nary a thought, who reach to grab something from the cashier, and who charge the wrong way down aisles, closely brushing past anyone there. As Mark Twain noted, most of the problems in the world aren't actually created by evil people, but by people who just don't care.

Weekends tend to be the time that those who have been seriously self-isolating tend to come out of hibernation and scuttle to the store for supplies. Peering around with frightened eyes and not paying attention to the changes that have been taking place in grocery (and other) stores over the past weeks. (Please, a little logic, guys. If you are concerned about the numbers of people you might encounter in the stores, the weekend is not the time to come.) Recently, possibly because of the reopening, there were a number of people who were even more cavalier about the new standards of shopping behaviour than before. I saw one fellow (in his twenties) on a skateboard. (In a grocery store? Really? You think that's appropriate? Even in normal times?) One person had his shopping cart on one side of the aisle and was wandering around on the other side, leaving a maximum of about 18 in. of space for anyone to get by him. I don't know what he was looking for, but he obviously never found it. I waited behind this unconventional but unpredictable barrier for about two minutes while he meandered. Finally, he looked at me and asked, 'You didn't want to pass me?' and then walked off, laughing derisively at me.

So I'm in the mall, checking out where everyone (anyone?) is, and swinging wide around corners so that I'm not going to suddenly be too close to somebody coming the other way, when the thought hits me that those movies and TV shows where the attacker suddenly jumps out from behind a door, or around a corner, are going to seem pretty silly for a while.

Which also reminded me that, these days, movies and TV shows where the gang gets together prompt an immediate emotional reaction of 'You're too close together!' (Gangster movies, where rival gangs 'meet' in a huge warehouse, and everyone stays a paranoid distance from each other, with hands on guns, prompts a 'that's more like it' response.)

Physical distancing (which is a more accurate, but less catchy, phrase than the more widely repeated 'social distancing') is one of the actions that we can take that are known to reduce the transmission of the virus. It's a reasonable precaution, it doesn't cost anything (except in terms of floor space), and it's not too hard to figure out. You can use it almost anywhere, including chance encounters and conversations on the street.

It does lead to some interesting oddities. I saw a 'Stay 2 m away' sticker on some construction equipment, obviously mandated by WorkSafeBC. Trouble is, it was on an asphalt spreader, making me wonder why anyone thought I had to be warned to stay two metres away from something like that.

Cryptography (Yes, Really)

7

As I mentioned, when I first thought about structuring this material based on the security domains, my immediate thought was that cryptography would be a non-issue. I was wrong again. It was that same day on which I found out about the Decentralized Privacy-Preserving Proximity Tracing (DP-3 T) and related protocols in regard to contact tracing.

As we discussed in Chapter 1 on confidentiality, integrity, and availability (CIA), the DP-3 T protocol is not the only version of contact tracing that is available. It is the type of protocol that most privacy experts are promoting. It is useful for contact tracing. At the same time, surveillance and abuse of the contact data base is prevented. Here is a rough outline of the basic privacy protecting concepts.

Each day, or at some other regular preset period, my phone generates a new number, also known as a beacon, which is similar to a nonce or salt. It is simply a piece of data, preferably random and long enough so that someone else is unlikely to have the same random number. My phone is regularly broadcasting that number. Just as yours is broadcasting your number. You could just keep the same number for a longer period of time, but the longer it is being used and broadcast, the more possible it is for someone to start tracking you by using that number. You could also choose or generate a new beacon every time you broadcast. There are three reasons you might not want to: (1) generating pseudo-random numbers is always a non-trivial task, (2) you will have to store that much more data, and (3) using a new beacon for every broadcast means you can calculate the duration of contact (which is important to the risk of infection) only if you also store at least the time the beacon was sent and transmit that data to the database.

When you and I are both at the same locale (or at least close enough for Bluetooth to pick up the broadcasts), your phone records my number for that day, possibly along with when and where it was 'seen' or 'heard'. There are going to be variations in the implementation of the protocol, depending upon how privacy-emphasizing you want to make it. It will still work with just random numbers, and obviously including time and location provides more chance of a breach of privacy but also allows more information to be given about where and when you might have been at risk.

If I subsequently test positive for infection, I upload my numbers (but not my name or any other location history that's not part of the app) for the past two weeks to a public database. Daily, your phone checks the public database to see if any of the uploaded numbers is in its local database. If it finds my number, it tells you that at some point in the past two weeks you were close enough to an infected person to get infected or possibly that at 3 PM on Wednesday you were within Bluetooth range of a potentially contagious person at McDonald's.

So, each day, everyone's phone downloads 88,000 new numbers (the worldwide total of new daily cases yesterday: the number will vary) and compares the list against the numbers it has collected over the past 14 days.

Unlike a centralized approach, this does not require big horsepower to centrally crunch large volumes of data and there is no central cache of personally identifiable information. Of course, it also means that the authorities do not automatically get a list of names for contact tracing (unless you turn yourself in).

As I mentioned earlier, there are variations in regard to this basic idea. As noted, more or less data can be sent to the central database. Sometimes, the number pseudo-randomly generated is itself hidden further by hashing which hides the actual data but can be used for comparison. If this area interests you, there are a number of resources for further study. Possibly the first is the paper which gives the technical details of the DP-3T model, which allows for low-cost, decentralized, and hybrid models for proximity tracing. It also details the privacy risks of all three versions:

https://github.com/DP-3T/documents/blob/master/DP3T%20White%20Paper.pdf.

Apple and Google have agreed on a common protocol and provided an API (application programming interface) which can be used by other developers to create consistent and somewhat interoperable apps. (Interestingly, their beacon, called a proximity identifier, has a lifetime of 15 minutes.) An amalgamation of comments on their work can be found at:

http://catless.ncl.ac.uk/Risks/31/67#subj2.

Application Security

8

Many of you will wonder if, just like cryptography, there are very few aspects of applications security where the pandemic and related issues can point out any lessons. Well, it turns out that there are a few.

Malware tends to get lumped in with the application security domain. Since that's where I started my security career, let's begin there. Along the way, we'll pass a number of other application security factors.

A COMPUTER VIRUS EXPERT LOOKS AT COVID-19

First off, let me say that, while 'virus' was and is a reasonably good choice as a term for replicating malware, it doesn't do to push the analogy too far. A computer—any computer, even a supercomputer—is a fairly simple entity in comparison with the complexity of the human body. And it's easy to say whether or not a computer is infected with a computer virus. It's pretty quantum. The computer is either infected or not. Either a computer virus is running in memory or it's not.

When I worked in the isolation ward and in industrial first aid, I learned a lot of things that later pointed out just how different biological and computer viruses were. And, when you study the various fields of science, which I did, you can analyze some of the factors that determine how viruses work.

In comparison with a computer, any body is more akin to, well, the Internet itself—a network of billions of computers (all the cells in your body), any one of which may or may not be infected.

A computer virus is just code. I have several thousand computer viruses in the office with me. Hundreds of them are on each of the computers I have.

They are of almost no risk to anyone since they are all either on floppy disks (those are of no risk to anyone who doesn't have a floppy disk drive any longer) or in 'zoo' directories. They aren't going to execute. They won't replicate unless I copy them somewhere. (No, don't ask. We old malware researchers are funny that way.)

A biological virus is alive. Actually, get a few microbiologists in a room together and making that statement is a good way to start an argument. There are a large number of factors that we generally consider necessary for life and that viruses don't have. But we can say that viruses are, at some point, viable and will replicate (under the right conditions) and, at another point, are not viable and won't replicate.

It's rather difficult to say that a person (a body) is infected or not. I probably have some rhinovirus in me somewhere, but I don't (at the moment) have a cold (that I know of). I probably have some flu virus (viruses?) in me somewhere, but I don't have the flu. There is a progression in most virus infections. A virus lands on you or you inhale or ingest one. (Actually, it's probably more than one 'copy' of the same virus. Infectious disease people talk about viral 'load' in reference to the number of viruses that you need to infect, that you have, or that you shed while you are infectious.) Your body has defences that are running all the time to fight off viruses, bacteria, parasites, and other things that shouldn't be in your body. But if there are enough copies of the virus, they may either get past or overwhelm your defences and begin to replicate.

At that point, you probably can be said to be infected, but you probably don't know it yet. The virus is attacking and spreading in your body but not to the point of causing symptoms yet. That is why you can be infected, and infectious, before you realize it.

The virus replicates by inserting its own genetic material into one of your cells and forcing the cell to reproduce it (generally destroying your cell in the process). (The genetic material of CoVID-19 is RNA rather than DNA, but in highly simplified terms, since we use RNA in the process of recreating our own DNA, this is not a problem. For CoVID-19. It is kind of a problem for us.) Viruses tend to prefer certain types of cells. CoVID-19 prefers lung tissue (among other types). Once a virus has started to reproduce on a large scale in your body, the fact that you are losing some of your cells, and the fight that your defences are making against the virus, produces symptoms. At this point, you are infected, and infectious, and probably know something is wrong.

Your defences have some generic ways to identify and fight off intruders. (These are akin to the change detection or activity monitoring types of computer antivirus programs.) But, when an infection actually takes hold, your body's defences learn how to recognize and target the specific infection. This process often involves antibodies. (This is similar to computer virus signature scanning types of antiviral programs.) (We'll come back to antibodies.) These defences may initially create additional symptoms or make the existing symptoms worse, but eventually they will build up and overwhelm the specific virus, drive it away—well, not completely, but to a very low level—and cure you. If the infection (or your own body's response to it) doesn't kill you first. As your defences are getting the better of the virus, you are still somewhat infected, still shedding copies of the virus, and therefore still infectious, but your symptoms are disappearing and you are feeling better.

Recently, research has discovered that the pandemic causing coronavirus is unique in short-circuiting the safest way our immune system kills off a virus. Interferon is actually not a single drug, but a family of proteins produced by the body's immune system in response to an invading viral infection. The proteins signal nearby cells to build up their defences against invasion, so interferon doesn't actually attack the virus but, by raising defences, interferes with the virus's ability to copy itself by using cell mechanisms. It is therefore the safest way for the body to deal with a virus. SARS-CoV-2 seems to somehow avoid detection by the body's defences that would normally trigger production of interferon. One of the scientists involved said that SARS-CoV-2 was like a stealth virus, which is really a very close approximation to the operation of the 'stealth' category of computer viruses, which in various ways mess with the indicators that we normally check in a computer to find out if a computer virus is present.

In people with CoVID-19, and in animals infected in laboratory studies, doctors and scientists say it seems like the natural interferon isn't activated the way it should be by the SARS-CoV-2 virus. When cells get infected, they usually prompt defences for the cell itself, and those around it, via interferon, which appears to be an alert for the immune system's first response. In addition, there is a triggering of a longer-term response by releasing proteins called cytokines. Unfortunately, overproduction of cytokines can create conditions known as 'cytokine storms' which can create more damage than the original disease itself. Most viruses block both of those roles. What makes SARS-Cov-2 unique is that it blocks the 'call-to-arms' function from interferon only.

I've mentioned the issue of viruses being alive versus being viable. CoVID-19 seems to need to be wet to be viable. It travels between people in drops of water or mucus (in very small drops, so we call them droplets). Without water the virus itself can't exist (or at least isn't viable). Some viruses can exist as a single virus with no water that can be breathed out and hang in the air for some time, bouncing between air molecules. We call them aerosols (and there are other types of small particles that hang in the air that we call aerosols), but CoVID-19 doesn't seem to be able to do this. (Sometimes, people say that coughing aerosolizes your saliva, but the droplets with water are much bigger than true aerosols.) The droplets have to be big enough to contain water for the CoVID-19 virus to be viable and infectious and that means that the droplets are heavy and therefore fall out of the air fairly quickly. They can't travel very far from the person who produced them. (This is where the 'six feet', 'two metres', and 'fingertip to fingertip' rules come from.)

The size of the droplets, and the inability to hang in the air, is why masks aren't very effective at preventing people from becoming infected with the virus (although they do help in some specific and dangerous situations where you are encountering a number of people with a high viral load who are coughing up a large number of droplets). Masks are somewhat more effective at preventing people who are sick from spreading infections since the masks, even just dust masks, catch the droplets. If you catch the virus, you probably won't breathe it in. You will probably touch a surface (any surface, even the surface of yourself or another person) where a droplet landed and then touch the mucus membranes of your eyes, nose, or mouth, which are nice and moist and which SARS-CoV-2 really likes. And the virus remains viable. And infects. (Are your eyes getting itchy just thinking about this? When was the last time you touched your eyes because they felt itchy? You touch your face a lot more than you realize. This is why constant handwashing (with soap) or hand-sanitizing is important. The outer envelope of a coronavirus is mostly a layer of fat, and if you have studied chemistry, it easy to see why coronaviruses really don't like soap or alcohol.)

TESTING

You may have heard that CoVID-19 can be detected in the air hours after an infected person has been there. You may have heard that CoVID-19 can be detected on surfaces up to three days after an infected person has been there (depending on the type of surface). There is a difference between 'can be detected' and 'is viable'. Remember that our current tests for CoVID-19 are checking for strings of the virus's RNA in the same way that computer antivirus programs check for strings of code that are unique to the computer virus. The virus or fragments of the virus (even if not wet or viable) can hang in the air or be on surfaces and be detected by RNA tests long after it has ceased to be viable and infectious.

There is another type of test involving the antibodies we spoke of earlier. Serology tests will not detect the virus directly but will detect whether someone has been sufficiently exposed to the virus to develop specific defences, such as antibodies, to the virus. This would indicate that a person has had the virus (and then recovered) regardless of whether they demonstrated any symptoms.

Serology tests will tell us other things about the virus and how it spreads, particularly about how many people in a given population become infected.

Testing is important in all areas of applications security. The two types of coronavirus tests point out an important lesson: different styles of tests very often give you wildly different information about a situation. They are also subject to completely different types of errors. In the coronavirus pandemic, the RNA tests tell you that you do have the virus right now. You are infected. That is, you are infected if the string of RNA that we are testing for does uniquely identify this particular virus. If the string is not carefully chosen, we may be identifying all kinds of things that have similar strings and functions to the virus we are really interested in. If the RNA string says you don't have the virus, then you don't have the virus. If, that is, the virus has not mutated recently, right in the middle of the string we are using for identification.

The serology tests, as noted, indicate whether you have had the virus at some point. They don't say whether you have an active infection right now. They may also point out that you were vaccinated or have had some similar virus and that you have antibodies to that other virus, which may or may not provide some protection against SARS-CoV-2.

Recently, a certain national leader has directed that testing be reduced so that the number of new cases of the disease will be reduced. This is, of course, flatly ridiculous. Testing does not cause problems; it just reveals existing problems. And the lack of testing doesn't prevent problems; it only blinds you to the scope of the problem.

I am reminded of a situation where sales and marketing was supposed to carry out virus scans before they installed our software product. They had previously been using an inferior testing program and I mandated that they use a more accurate test. At one point, a machine was brought in as a problem. First step in my process was to scan the machine, and, sure enough, it was infected.

'Did you scan it?'

'Yes'.

'Did you use the right scanner?'

'Well, no, we used the old one'.

'Why did you use the old scanner, when I've specified that you have to use the new one?'

'Well, when we use the one you told us to, it finds viruses'.

We are security professionals. We deal with risk. We know that risk always involves probability. A biological infection situation is not quantum. It is not 'If you leave the house, you will get infected'. It is 'If you leave the house, there is a higher likelihood you will become infected'. Biological virus infection involves proximity to an infected person, time of exposure, that person's viral load, number of proximal contacts, and a number of other factors. And all of the various factors involve probabilities.

The probabilities can add up. If you pass someone on the street or in a store, there is maybe a one in a million chance you will become infected. (Don't quote me on the 'million'. It's just for this example.) That isn't big. (If you own a pool, the chance that you will die by drowning is twice that level, but many people accept the risk.) We could avoid that risk by not going out, but then there is a risk we could starve to death, so we have to calculate and balance those risks. But if we encounter ten people at that store, those risks add up, so now we are at one in a hundred thousand. And if we go to ten stores, then we go to one in ten thousand. And if we keep that up for ten days, then we go to one in a thousand, and if we keep it up for three months, we are at 1 %. Which starts to sound like it might be a bit dangerous when the impact is that we might die.

So we have rules. But the rules are based on probabilities. It's not that at six feet you are safe but at five foot six inches you will be infected, but rather that it is unlikely that droplets will easily jump six feet. They will more easily jump three feet, although it's still not guaranteed. Rinsing your hands with water will get rid of 80% of germs on your hands. Washing with soap and water for 20 seconds and the proper process will get rid of 99.9% of germs. However, if you are pretty sure that you've touched something that might be dangerous but you can't, right now, wash thoroughly but you can, right now, rinse your hands, then rinsing your hands right now is better than doing nothing. (Although you should make sure you wash your hands thoroughly as soon as you can.) All of our 'six feet'. 'wash hands'. and 'don't congregate' rules are risk mitigation.

(No, for those students of risk management, there is no risk transfer in this scenario.)

And remember the tests that can't tell the difference between viable and dead viruses and the studies that say the virus can live on surfaces for three days (if metal or plastic) or four hours (if copper or cardboard or steel but in direct sunlight)? It's not that all the virus copies stay alive for 72 hours and then die on the 73rd. Copies of the virus are dying all the time, and after a certain number of hours, half of them are dead, and after that same number of hours, half of the remaining ones are dead, and all that time the viral load is going down and the probability that there will be enough copies of the virus to actually infect you is decreasing.

So you calculate the risks and assess them the same way you calculate that it is unlikely you will be stabbed to death if you go to a party. (Wait. You were at a party? During the CoVID-19 crisis? What kind of risk management decision is that?)

So what lessons have I learned, as a virus researcher, that **this** virus has reminded me about?

As I have mentioned before, I have seen (over and over again) that when people don't understand something, and are afraid of it, they tend to freeze. They tend to build false rationales to justify avoiding the issue or not doing anything. The first thing you have to do is raise awareness and educate before you can help people to realize they need to follow some guidance in order to stay safe. If they don't freeze, they tend to build the most wildly inaccurate mental models of the situation, based on little or no evidence at all, and this misinformation needs to be challenged before the reality can be taught.

I well remember the day when, talking to a long-time friend who was an educated and experienced businessman and CEO of a very large enterprise, I mentioned our research into virus writers. He was absolutely shocked. 'People *write* these things?' he said. Yes, of course. How did he think they came about? 'I thought they just sort of ... developed'. he said, rather miserably. I did once do a calculation about how long it would take for a computer virus to 'just develop'. I came up with an answer that was longer than the lifetime of the universe. I may be wrong because a more famous virus researcher did the same thing and came up with a shorter answer. But we agree that the answer is more than millions of years.

In security, if you are successful, people will think you have overreacted. The Michelangelo virus in 1992 was the first time that we in the computer virus research community were successfully able to warn people and raise awareness of a pending virus threat. We got hundreds of thousands of people around the world to check their computers, and they found millions of copies of the virus. Before the virus was due to trigger (because of the leap year) on March 5. (It had been around for over a year before that.) And on March 6, all the news stories were not about how well we had done, but about how the warning and 'panic' had been a hoax.

We are all in this together. A computer virus researcher sees security from the inside out. Most security people grew up with the bastion mentality: we are on the inside and the bad guys are on the outside. We build walls and protections and try to protect from attacks from the outside. If a bad guy is attacking you, it isn't my problem. As a matter of fact, it's a good thing, from my perspective, because while he is attacking you, he is too busy to be attacking me.

A computer virus researcher realizes that this viewpoint is very badly flawed. Attacks can come from within as well as without. If someone attacks you with a computer virus, then it is a concern for me because, if you get infected, you will be sending out copies of the virus as well. An attack on one person raises the threat level for the entire community.

For years, therefore, I have been emphasizing the idea of a kind of public health security awareness campaign. I have put it to major corporations that they should be promoting and paying for security awareness campaigns, not just for their employees but for the general public. As well as the public relations and goodwill value for the enterprise, it is to the corporation's advantage.

The greater the knowledge of computer security there is in the general population, the lower the threat to the company—any company.

So it is with CoVID-19. We are all in this together. Many have noted that we are all in the same storm, but we are not in the same boat. My threat level is not the same as my granddaughter's; she is young, female, and fit and has no underlying health conditions. My threat level is also not the same as that of my friend in the United States, who has said there is no point in going to the doctor since, even if diagnosed, she would not be able to do anything about it since she doesn't have medical insurance. (As a Canadian, I find that appalling but unfortunately completely logical.) But we are all in the same storm. We follow directives to stay two metres apart, wash our hands, avoid crowds and concerts, and not visit old relatives, not only because of our personal danger of getting infected but because every individual infection adds to the total risk for our immediate communities and society as a whole. We bend the curve (and not the rules) even though the same total number of people may die. We do this because the overwhelming collapse of the medical system and infrastructure as a whole may not affect me personally but is a bad thing overall. Not hoarding N95 masks and other personal protective equipment, thus leaving it for those who really need it, keeps nurses and other medical personnel safe and thus keeps us all safer.

TOILET PAPER (RECIDIVUS)

Malware is, of course, an issue that we must always take seriously. So, in view of the panic-buying–induced shortage of toilet paper, it was ironic that—shortly before the pandemic was officially declared but after toilet paper had disappeared from most store shelves—three major pulp and paper mills in British Columbia had to reduce production because the computers that normally managed the processes were found to have been compromised by malware. Important data had been corrupted, and email communication within the company was disrupted. The automated system for ordering was also shut down.

So this means that everyone who has stockpiled toilet paper should safely dispose of it because it may have been infected with a virus. (The scary thing about making that joke is that I have to tell people IT IS JUST A JOKE, and computer viruses can't jump from computers to people or inanimate objects, and some people will STILL believe it and take it seriously.)

More seriously, you have probably seen screenshots, on the news and in reports, of the CoVID-19 dashboard and map created by Johns Hopkins

University. This map, very valuable and widely used, was 'scraped' and reposted at a variety of sites around the net. Most of those sites served up spyware and other malware. (The original Johns Hopkins site is fine.) This misuse of a legitimate resource tends to happen a lot around disasters and crises. Many sites were also created to fraudulently request donations, supposedly to be given to victims and those out of work. Sadly, there are always those willing to profit off the misery of others.

Oh, by the way, during this crisis, both medical and economic, lots of people are trying new ways to make money. According to movies and TV shows, the quickest way to make a great deal of money (aside from just stealing it) is to trade stocks, bonds, and various financial derivatives. So, many people are trying their hand at stock trading, and lots of developers are selling systems and apps to help them. One such app was specifically aimed at novice investors just getting into the trading game. However, it had an unusual user interface that did not always display the complete state of the trader's financial situation (and does, in fact, make it look very much like an online game). It would, on occasion, display status after only part of a trade had been completed and, in this case, could show the trader as having lost huge amounts of money and being in significant debt. In one such case, the young trader committed suicide.

The user interface can be important.

Security
Operations

9

DO WHAT YOU HAVE BEEN TOLD

The future is in our hands, and we must continue to wash them.
                                                        –Dr. Bonnie Henry

You may have noticed, gentle reader, that I have, at various points in this book, dropped the adjuration to wash your hands into seemingly unrelated text. There is yet another lesson in my doing so: when teaching security awareness, remember that your primary model is advertising. Repetition, repetition, repetition. Drive the point home. Keep reminding people. In as many different ways as you can. Or just keep repeating the point.

Now go wash your hands.

There is yet another security lesson related to this one: the most important acts and principles are the simple ones. You don't need hydroxychloroquine, you don't really need remdesivir, you probably won't need a ventilator. But you do need to take simple precautions: so simple that many people probably think, 'It can't be that easy'. It can and it is and you need to do it. (Them?)

The World Health Organization is currently promoting Five Heroic Acts (which is rather brave of them since the phrase obviously pokes fun at the Chinese Communist Party with its humourless fondness for unimaginative and baldly descriptive naming conventions). There is absolutely no surprise in any of the 'acts'. In no particular order of importance, they are:

- Keep your distance. (Yup. The old six feet/two metres rule.)
- Sneeze into your elbow. (Which is only code for 'don't cough onto people' and, if you do cough into your hands, don't just ignore the fact that you've coughed germs onto your hands and then touch everything with virus-laden hands.)

- Don't touch your face. (You are going to touch stuff—door handles, elevator buttons, stray surfaces—with your hands and possibly pick up the virus. Don't then transfer it to your face.)
- Wash your hands. (Yeah, it is that simple. It also works.)
- Stay home. (It's a nice place. And safe.)

(By the way, even after the pandemic is finally over, these five acts would save many lives every single year by reducing the severity of flu season.)

MASKING THE COVID-19 PROBLEM

Which brings us, once again, to masks.

Masks have become an incredibly divisive issue, very surprisingly so, from my perspective. Of all the various items that I've posted on the Internet over the past few months, the one that got the most response was on masks. And the thing is, those responding didn't have any particular points they wanted to make about any errors or omissions I had made in my posting. They were making direct, personal attacks on me, an indication that they were objecting out of emotion rather than reason.

Possibly this has to do with the fact that CoVID-19 and the pandemic are huge, complicated, and messy situations. Most people aren't doctors, let alone epidemiologists, let alone virologists, let alone specialists in the coronavirus family. And the pandemic has created problems in business and industry and transport and supply chains and economics and governance and recreation and has had a huge impact on daily life. The CoVID-19 pandemic is an enormous and complicated disaster in many areas where most of us do not understand all the ramifications.

But masks are simple. We understand masks. (Or at least we think we do.)

The wearing of masks has become freighted with all kinds of issues beyond the medical. Masks already had significance in many Asian cultures, where large populations lived in cities or regions with poor air quality and high air particulate counts. Wearing masks was related in some sense to public health though not necessarily to disease and infection and was, in a way, accepting the necessity for industrial activity to promote the economy and accepting that the governmental and societal requirement for economic growth was more important than your need for clean air. Therefore, wearing a mask was again making a statement: that you were subordinating your needs to those of the community.

In Western cultures, the wearing of masks has cultural freight as well, though slightly more complicated. Prior to the pandemic, masks had been used primarily by protesters to make identification (and later arrest) more problematic. (Also as a minor protection against tear gas and other forms of crowd control.) Therefore, the statement about masks was more of a protest against state surveillance. But some protesters, noting that many of those wearing masks were agitators rather than devotees of the cause underlying that particular protest, came to believe that not wearing a mask was a statement that they had nothing to hide. (Which makes the issue of mask-wearing during the 'Black Lives Matter' protests all the more complex.)

In addition, in the United States, at the very least, masks have become freighted with even more baggage in regard to political affiliations and concerns. Wearing (or not wearing) a mask may display yet another statement: that of alliance with a particular political leader or ideology which may assert that the pandemic itself is either overblown or not an important issue in the grand scheme of things.

Masks aren't just about preventing infection anymore. Which provides another security lesson for us: sometimes the latest security buzz phrase or technology is more important politically than it is in terms of actual security.

Properly fitting and properly filtering face masks are important parts of medical personal protective equipment for keeping front-line medical staff safe if they are in areas or situations of high viral load. (Or, indeed, in many other situations where they may be encountering any number of infectious agents.)

Otherwise, having a piece of paper or fabric in front of your mouth does very little to keep you from becoming infected with the CoVID-19 virus.

Let's look at the 'evidence' for the benefit of masks. There seem to be two points of evidence.

The first piece of evidence is that nurses and medical techs wear masks. You can see them. They are the 'face' of the medical system and, these days, that face is covered with a mask. Obviously, masks are important.

'Obvious', as we say in mathematics, is what you claim when you can't prove it. You do see nurses, medical techs, and emergency first responders (on the news) and staff in intensive care units (on the news) wearing masks. You can't help but notice the masks. You don't notice gowns (changed between patients), gloves (changed between patients), face shields, and constant, everlasting handwashing. You also don't see the vastly higher probabilities that these people will encounter the virus, nor the fact that the gowns (changed between patients), gloves (changed between patients), and masks (changed between patients) are intended as much to protect you as the medical staff. (Nor the 'public relations' and 'social engineering' aspects of 'security theatre' intended to sooth fears in a time of poorly understood crisis. There are non-medical reasons to wear masks in some situations.)

The second piece of evidence is an 'observation' (one cannot call it a study) that some populations with a high incidence of mask-wearing have significantly lower transmission rates of the virus. (You cannot call the observation a 'study' since the sample size is very small. We are talking about whole countries and not just countries but 'countries with high rates of CoVID-19', which takes you down to a double handful. A 'double handful' is not a statistically valid sample set.) There are two additional (and easily observable) factors that may affect the transmission rate without recourse to the idea that masks prevent infection. The first is that masks are demonstrably effective at reducing the probability that those who have the virus (and 25% to 50% of those who are infected show few or no symptoms and don't know they are infected) will directly pass the virus to others. (Masks, of pretty much any kind, tend to vastly reduce the droplets breathed or spit out by those infected, simply by trapping the droplets as they come into contact with the fabric or paper of the mask.) The second factor is that those countries with low transmission rates also have authoritarian governments, which can quickly and effectively mandate that people must stay home and isolate themselves.

Recently, on the Dr. Bonnie Show (co-starring Adrian Dix and Nigel Howard), a reporter 'asked' a 'question' designed to give warning that he would be running a story on the dinner-time news hour that N95 masks were available, at various locations, and implying that the provincial government was not doing enough to find masks for medical personnel through local sources. Without even having to draw breath, Dr. Bonnie pointed out that N95 (and I myself have been guilty of being careless with this distinction) is not a type of mask but a type of filtering material. It is a filter that can trap or eliminate 95% of particles of more than 5 microns in size. Once you have that filtering material, you can use it in dust masks (which you can usually buy at any hardware store), face-fitting masks, medical masks (needing water-repellent coverings and possibly other capabilities), and various types of respirators (with or without built-in goggles).

With your own, homemade masks, you might be protecting others, but it's not highly probable. Yes, masks trap droplets, but that matters only if you are infected. Even if you live in the United States (as I write this, the world leader in cases and deaths), there is only one chance in one hundred that you have or have had the virus. And if you know you are infected, wearing a mask does nothing if you are alone at home. If you are infected, you should be home alone. What are you doing going out if you are infected? Do you want to kill people?

Okay. You wanna wear a mask when you go out? During the virus crisis, if you must go out, note that you might get coughed on or sneezed on, and since disinfecting fabric is much more difficult than cleaning flat surfaces, you should wear older clothing that can be discarded if necessary. (If you have old torn clothing that will not be missed, this is probably best.) Since face masks are in short supply, a scarf worn over the mouth, nose, and lower part of the face may offer some protection.

Masks are not magic. In my opinion, there is as-close-to-zero-as-makes-no-difference evidence that masks prevent normal people, in normal situations, from becoming infected. 'Magical thinking' will not help us in this virus crisis. And it may do an awful lot of harm.

There are reasons to wear masks. Workers in medical situations are likely to encounter, without warning, persons with high viral loads. Workers in medical situations, likely dealing with all kinds of people with immune systems that are compromised for various reasons, should take every precaution against transmission of the virus they may not yet be aware they are carrying. If you are forced into a situation where you cannot guarantee physical distancing (like public transport of various kinds), a mask does not replace the better forms of protection but will give you some small amount of additional protection. And, in some situations, social engineering comes into play again, and wearing a mask may help calm people you have to deal with, even if it provides no actually necessary protection against infection.

(By the way, if you are forced into close proximity and you don't have a mask available, breathe through your nose. The nose isn't perfect, but it is a rather remarkable filter, designed by millions of years of evolution. It is actually a little susceptible to infection by SARS-CoV-2, but it's less susceptible than your mouth, and breathing out through your nose means you produce far less in the way of droplets to spread the virus if you have it.)

Now go wash your hands.

I PROTEST THE CROWDS

As I write, the Black Lives Matter protests are in full swing across the US and in other countries as well. We definitely need police, particularly in our large, complex, and inclusive society. Police are probably the number-one 'first responders' and yet they don't get the credit and social aura that firefighters do and nurses have in this crisis. The calls to 'defund police', especially in the midst of this crisis, are remarkably short-sighted or, at best, simplified to the point of absurdity. We have rather blindly allowed the job of the police officer to expand enormously as society has become ever more complex, and we have not supported the law enforcement departments and agencies with the resources to effectively address this ever-increasing task. But the statistics do not lie. The numbers demonstrate that it is also true that many law enforcement agencies (and we as a whole society) have allowed racially based bias to contaminate the direct implementation of the vital and crucial role that police play in our communities. Most police officers are sincere in their desire to universally and even-handedly 'serve and protect' the members of their community, but constant repetition (remember repetition?) of even the slightest unchallenged slurs does have an insidious and pernicious effect on even the most well intentioned. The protesters, and the protests, do have an important point to make.

Yet it shouldn't be that the risks to the community of racial bias require addressing by yet another risk to the community. And I am not talking about violence and looting. I am talking about the inherent risk, during a pandemic, of gathering large numbers of people together. Yes, there were occasionally attempts to have the crowds keep 'social distance'. But it's extremely difficult to get an undefined and undifferentiated (which, I suppose, is kind of the point) mass of people to keep the necessary physical distance from each other. (However, I would make the point that, if you can get everyone to stay at least two metres apart, your 'protest' will take up impressively larger street space. Just saying.) Yes, a lot of the people in the protests wore masks, possibly as much to make identification difficult as for public health reasons. (But you'll also note that masks, in general social situations, are one of the least effective protections and not one of the World Health Organization's Five Heroic Acts.) I'm a little disappointed that, out of all the various protests that have been mounted, almost none managed to find creative new ways to draw attention to the cause in a safer (to the participants and the community) manner.

In pursuit of the concept of convincing people not to join crowds of protesters because of the pandemic, the use of tear gas and other 'pain gas' crowd control measures is remarkably stupid. Tear gas and other gas irritants are known to cause damage to tissues in the respiratory tracts, which is known to be a factor in CoVID-19 transmission and, in addition, is known to depress immune response.

Think beyond the immediate. Think through the implications of what you are doing, in both public health and security. As lockdown and self-isolation rules were being lifted (or even slightly before), a number of people wanted to 'expand their bubble'. What is the harm in allowing one more person or one more family into your household 'bubble' of people who have been self-isolating together? It sounds so reasonable. Just one more person. Somebody you haven't seen for a while. A family member. Let's go and visit Granma at the old folks' home. What can it hurt?

As I've mentioned, doing a partial job on security can be worse than doing none at all. It doesn't matter that you've got terrifically good locks on your door if you have an open window right beside it. It doesn't matter if you've got the best firewall in the world if the admin password is weak. In public health, having just one more contact seems innocuous. But if everyone has 'just one more contact', it opens up a huge network of interconnections that allow transmission and infection through the entire community. In relation to CoVID-19, if you let just one more person into your bubble, you also let in every 'just one more person' they have had contact with and every 'just one more person' those people have had contact with, and so on and so forth.

CONTROLS

We in the security field are 'controls' freaks. Not control freaks (although some say we are that too). We are continually interested in controls that enable us to safeguard systems and data.

Controls come in a number of forms. They may be administrative, such as policies and standards. Some controls are physical, as walls and removable media are. Increasingly, many controls are technical (or logical) measures like encryption and antivirus scanning. In planning and considering the types of controls we have, their effectiveness, and new ones we may need, we find it helpful to categorize controls into these different types. This tripartite arrangement of security controls has developed from the normal divisions of responsibility in business: management, physical plant, and operations (technical, in the case of information systems).

We divide controls into other classes as well. Corrective controls are applied when others have failed (helping to build our defence-in-depth strategies), directive controls provide guidance, deterrent controls use social pressures to reduce threats from human attackers, detective controls determine that a breach has taken place (and may gather or analyze information about it), preventive controls reduce our vulnerability to threats, and recovery controls assist us to resume operations after an incident. This six-way partition of security actions has its roots in military and law enforcement studies. Compensating controls may be a seventh class: they are very similar to corrective, although they also include provisions for the normally expected run of errors.

As we create security controls, the finer the gradations and the more subtle the distinctions we can make between safeguards, the better our analysis of our total security posture. Having two taxonomies of controls, though, frequently confuses both students of security and security practitioners, who want to know whether a preventive control is supposed to fit under administrative or physical controls and questions of a similar nature.

In fact, the two classifications are orthogonal. Trying to fit corrective, directive, detective, deterrent, preventive, and recovery divisions into the administrative/technical/physical structure does not fit, nor should it. The approaches, philosophies, and intents are quite distinct in the two different formations. Some time back, I created a 'controls matrix' to illustrate this issue for students of security, and it turned out to be a remarkably useful security framework for analyzing your existing security strategies.

At any rate, I was intrigued to find, in the various discussions about the CoVID-19 pandemic, that the administrative/technical/physical division of security controls has shown up. The names are slightly different. Technical controls, in a medical situation, tend to be treatments of various types, whether those are preventive treatments (like gloves or vaccines) or actual treatments for the disease. Physical controls, in medical terms, seem to be called engineering controls, like the Plexiglas shields being put up in retail establishments. Administrative controls are, as they are in security, the policies and tend to be referred to just as 'rules' or 'directives'.

INFORMATION SECURITY FOR WORK-FROM-HOMERS

During the pandemic, one of the many things I saw posted was a page of advice on how to work from home. It was the standard, pedestrian, banal, 'get yourself organized' advice that everyone always posts about working from home. I've been an 'independent consultant' for a few decades. I also wrote four (or six, depending on how you count them) books, which involves a lot of working from home. Here are some of my tips on working from home:

You may not have enough distractions at home. You know home. You are intimately familiar with everything in it. Familiarity breeds contempt.

If you have access to a young child (or possibly someone with Down's syndrome or dementia), they view the world in a different way. This can be extremely valuable to you. (Many people will think I am joking about this. Absolutely not.) If you work at it, you can take this experience with alternative views and, with practice, begin to look at the world differently yourself.

You are unsupervised at home. This can be a great advantage. It is possible that you may simply slack off. However, it is also possible that you can look at what you are doing, and how, and figure out what is actually worth doing.

When you work from home, you can develop an 'always on' mindset. This may come about in two ways. You may be a natural workaholic and just get 'stuck in'. Or you may think that what you are doing is actually important. This can create a serious and damaging delusion. Once again, look at what you are doing and try to determine whether it is actually worth your effort and attention.

You may feel lonely and isolated at home. Especially if you don't have a spouse or children. Then again, you have the Internet. Good grief! You can talk to anyone in the world! About all kinds of topics! How can you feel lonely and isolated?

If you have a spouse or kids, you may feel that you can't be with them. You may feel that you have to lock yourself in your home office for eight hours a day and not talk to anyone. That's ridiculous. Talk to them. Play with your kids. Go for walks. (Stay away from anyone on the street, though.) Make multiple short blocks of time to lock yourself in your office if you must. You will be much more productive, in one hour, after you've spent some time actually talking to your spouse or playing with your kids. (And, again, how much of what you are doing at 'work' actually needs to be done?)

You may need to be motivated. Why? Isn't your work important or interesting enough? If that's so, it should be motivation to find a different job. (You can search for one on the Internet.)

Get distracted. Look out the window. Talk to someone on the Internet. Talk to your kids. Clean up your desk or office. You can find some interesting new ideas.

Don't procrastinate. (At least, not right away.) Do something. Anything, really. If a job looks too big and daunting, do something related to it to chip away at it.

Create a schedule. It does help. But create it. Don't just impose it on you and those around you. Find something that works for everyone.

TESTING OPERATIONS

Security operations can be simple. But they can also become very complicated. A number of factors need to be assessed in any situation to make sure your operations are going as planned.

The province of BC (British Columbia, where I live), in Canada, as previously noted, has had relatively low numbers overall. However, during July and August of 2020, there was a surge in cases. Even though BC has had a decent system for testing, the additional stress showed up a number of weaknesses and points out the types of things you should consider in any security operation. It also demonstrates that there are a number of limiting factors that should be considered in any operation, not just the maximum capacity of the system at the centre of it. The lab capacity for CoVID-19 tests in BC is 8,000 tests per day and, even with the surge, nowhere near that many are being carried out. But, as this experience indicates, a number of issues can impede access to a test.

Recently, a number of reports have shown that the elderly, despite showing no symptoms other than fatigue, can very rapidly move from fatigue to pneumonia and then death. So, even though I had no symptoms other than fatigue, my doctor wanted me to get tested. There is one (1) testing location on the North Shore. There is no available street parking around it. There is no street parking longer than one hour for three blocks in any direction. There is some pay parking a few blocks away, but you have to pay in advance (and thus know how long you are going to need to park). (I could take the bus, of course, but these days you have to factor that as an extra risk.)

I went to the testing location and parked in a one-hour spot. Walked over. There was a serious lack of signage outside or at the doors of the building. None. (Operations lesson: extra information and signage are always good things.) Inside I did see a line-up of people, which I took to be a positive indication, so I joined them. Then, eventually, a staffer came along and told me that (a) the line-up actually started outside (I had noticed a few people seemingly milling about) and (b) the wait time was three hours. I asked what time the testing centre opened. She said 8 AM but that people were lined up inside, even though they weren't supposed to, when she got there at 7:30 AM. I left.

Starting with the BC Centre for Disease Control (BCCDC) Website, I eventually found some information about the testing centre. It wasn't easy. (Operations lesson: Website design, interface design, and information flow consideration are all non-trivial tasks.) The front page of the BCCDC Website right now gives CoVID info, but the first time is about the self-assessment tool which the government provides (which has no testing information) and the second is CoVID information for the public. Far down on that page it mentions testing. The testing information link gives you a list of symptoms to decide whether you should get tested. Then it mentions travel and twice more suggests the self-assessment tool. Finally, there is a terse link labelled 'Collection centre finder'. (Operations lesson: consider what people might want to be most concerned to know.)

(There is also an automated bot, the 'BCCDC COVID-19 Digital Assistant', which you can 'ask' a question. When I typed in 'north shore testing centre', I got a canned piece of info again talking about symptoms but finally giving a link to the testing information page.)

That collection centre finder link brings up a map where you can zoom in and finally find that, yes, there is only one testing centre on the North Shore. (I first tried this on a smartphone, and the interface there is not easy. Operations lesson: note interface differences on different devices.) Clicking on the red dot provides some information, including the address and (finally!) times of opening. It's open from 8 AM to 10 PM, except 9 AM to 5 PM on Sunday.

There was also a link to a separate page for this specific location. Unfortunately, this page disagrees on times (8 AM to 9 PM, except 8 AM to 4 PM on Sunday). But it also gives a phone number. It also provides a handy 'Estimated Wait Time'. I was heartened to note that, after having been told the wait time was 3 hours at 1:30 PM, by 2:15 PM the wait time was showing 2.5 hours, and I was even more heartened to note that by 2:40 PM it was only 2 hours! When I checked again at 4:30, it was showing a confusing 'At Capacity'. However, building on the hope that the previous reductions in wait time had engendered, I went back down and actually found a parking spot (available only after 5 PM) that I could stay in until 10 PM if necessary. I walked over to the testing centre.

Here, I found the first actual sign from the testing centre, topped with the sign 'CLOSED'. 'At Capacity' apparently means that the line is long enough that they are not letting anyone else join it for that day. (So those who were there at 4:30 PM and did get in were facing a five-and-a-half-hour wait in line.) (The sign did also confirm that the actual hours were/are 8 AM to 10 PM and 9 AM to 5 PM.) (Operations lesson: consider that your terse codes might not clearly communicate the actual situation to those reading them.)

The next day, the specific Website with the wait time estimate showed a four-hour wait time and it only got longer.

Meanwhile, the collection centres map showed that there are five centres in Vancouver. One, downtown, showed a three-hour wait. Another is for the Down Town East Side (an area with a problematic homeless population) and has no Website and so no time estimate. Another, at Commercial Drive, requires an appointment and has no Website. There is one at Women's and Children's Hospital that has limitations and no Website. There is a drive-in at 33rd and Heather that has no Website. There is a drive-in in Burnaby that has a Website with some times and confusing directions but no time estimates. (Operations lesson: consistency is important.)

As noted, I did find a couple of drive-in centres. However, local news noted that the drive-in centres reach 'capacity' within about 45 minutes of opening. I'm not sure how much of a health benefit there is in waiting six hours in a line-up of idling cars.

The next morning, I got up early and drove down to the centre before it opened. A few blocks away, I found two-hour street parking, which didn't come into effect until 9 AM, so I figured I was good until 11 AM. I took a folding chair with me to have something to sit on in the waiting line.

I arrived at 7:30 AM. At that time, I was second in line outside the foyer, but at 8 AM when the office opened and the security guards came on duty, one of them cleared some of the people inside to outside. (I later found out that the inside line-up was eight people.) At that point, I was about sixth in line outside. At 8 AM, the Website showed a ten-minute wait.

At 8:30 AM, it showed a two-hour wait. (I had moved about three spaces in line by that time.)

This testing centre is co-located with an urgent primary care centre (UPCC). The line-up is just for the CoVID-19 testing. UPCC patients bypass the line, and staff (or security) send them directly to the office. CoVID-19 testing people get taken in when the UPCC office is not too busy with other people.

Up in the UPCC clinic, things are a bit more confused. There is a wait for registration. Once registered, you have a bit more freedom. (By the time I was registered, it was 10:30 AM, so I talked to the desk and then went and moved my car so I wouldn't overstay the parking limit.) You wait for a while and then see triage. Everybody goes through triage, UPCC and testing alike. Once you're through triage, there is yet another waiting area for CoVID-19 testing: they 'batch' the CoVID-19 tests in groups of four. (This is part of the reason the line-ups in the foyer and outside are inconsistent in movement. Also, UPCC cases may be given priority, depending upon severity.) Once the CoVID-19 group area is full, a 'batch' takes only about ten minutes.

I was done about four hours after I first arrived. I figured that the people at the end of the outside line, when I left (at 11:30 AM), were looking at about a six-hour wait. (The Website estimated wait time, at that point, was showing ten minutes, so obviously it applies only to the UPCC patients and not the CoVID testing. Operations lesson: be clear in your communications.)

The BC testing system has a number of options for letting you know your results. (My doctor had noted, when she asked me to get tested, that if the test was positive the testing centre would call me, and if it was negative, she would.) I registered (online) for notification via text message. I got my result (negative, thanks) about 25 hours after I was swabbed.

As noted, this is an experience with a relatively robust and well-planned system. However, as also noted, there are always ways to improve, particularly as situations change.

This experience also points out how important it is that we start planning (now, even before we have a vaccine) a distribution system and operation for whatever vaccine is developed. The plan may have to undergo some modification as we encounter special issues. For example, the vaccine candidates from Pfizer and Moderna apparently have pretty stringent refrigeration requirements.

Telecommunications and Networking Security

10

During the coronavirus/CoVID-19/SARS-CoV-2 crisis, a number of people have said that they will be doing events remotely, and I'm quite sure that even more will be jumping on the bandwagon.

In 1985, I ran the technical side of the World Logo Conference (WLC), which was the first ever fully integrated onsite/online conference. Subsequently, I worked as a data communications consultant in a wide variety of similar situations and events. I've done a number of online courses, Webinars, Webcasts, remote presentations, and other such events. That's thirty-five years of experience in the field. (Longer, really, because my previous experience was what placed me in the WLC job.)

Holding the onsite side of the hybrid conference in Paradise Valley, Squamish, British Columbia, was a non-trivial task. We used CompuServe, about three modems, lots of floppy disks for other machines (mostly reporting feedback from the onsite sessions), and rolls and rolls of printer paper tacked up to 'corkboards' along the walls. We had 75 participants onsite and 200 from around the world. I remember moderating one session at 5 AM. As far as I know, nobody has ever done anything like it since. I'd really like to attempt another hybrid conference to determine whether we can't do better with the technology that we now have available.

Warning: videoconferencing is not the same as having a face-to-face meeting. I'm sure that various psychology grad students are, at this very moment, doing tests to find out why this is so, in order to write theses and dissertations on the topic. We have trouble trusting discussions in videoconferencing because we look at the picture of the person on the screen, rather than at the Webcam, and thus appear to be avoiding eye contact. Eye contact is one of the body language cues we use to assess whether someone is trustworthy. We have become used to sitting around a large table and talking to each other. We have gotten used to various protocols regulating who is allowed to speak and when. The examples of how to do this can be used to teach network access protocols: there is the sequential 'round-robin' protocol (known in networking as polling), the 'talking stick' (token passing), or the cocktail party (carrier-sense multiple access with collision detection, or CSMA/CD). We are able to figure out, from body language and other clues, when somebody might actually be saying something important. We are able, in the long periods that someone is just burbling on, to hold side discussions with colleagues. We are able, when we are speaking, to figure out whether we have convinced someone or are holding their attention, again by subtle body language indicators.

With videoconferencing, almost all of this is lost. Most people don't understand the significance of lighting or camera placement and feel that, if they can be seen at all, it's fine. The pictures you see of people are thus from odd angles and with strange shadows, and people simply do not look like the ones you know. The pictures are generally small, and you lose a great deal of what body language normally tells you. In particular, since everyone is in their own frame and those cannot be arranged by seating preference or even your own choice, almost all normal indications of social dynamics and alliances are lost.

There is a condition jokingly referred to as 'Zoom fatigue'. It is no joke. It's quite real. Videoconferencing is not something we are used to, and it contributes real stress in an already stressful situation. There are a number of additional factors involved in planning a meeting, conference, or seminar via videoconferencing. Videoconferencing is not an easier way to meet: it requires much more preparation and background resources than an in-person meeting and some significant additional supports to make people feel comfortable and to provide for work to be done.

It's been surprising how arduous it is to raise interest in setting up or attending an online meeting or presentation. Someone else complained about it: 'Even CPE [Continuing Professional Education] won't get my co-workers to attend. I hear the response, "If I'm not being paid to attend, I won't attend". So much for learning, sharing knowledge, and networking'.

Part of the problem may be that everyone has become habituated to the low quality of 'Webinars'—thinly disguised sales pitches with almost zero information content—that vendors put out. Now we are seeing similar things happening to conferences: in order to maximize revenue/profit, conference organizers are signing up all the vendors they can, letting them give sales pitches as 'presentations' and putting no effort in (and leaving little time for) any actual presentations based on research or analysis.

There's also the fact that those of us in security, unlike a huge proportion of the population, (a) can work from home, (b) are working from home, and (c) in many cases are busier than ever (with fewer resources being provided by employers). I can see that—with everybody 'maxed out'—meetings, presentations, conferences, and Webinars would be a lower priority.

The fact that everyone is busy is bringing another problem to the fore. I had a meeting this week with other security leaders, and a good chunk of the first part was taken up with stories of burnout, stress, the need to take time off, the need to **force** people to take some time off to rebalance, and so forth. We probably need to take some care that we don't become too stressed. That can be difficult. Under stress, your analysis and decisions—including the assessment of how much stress you are under and how it is affecting you—may be affected.

These days, if you are doing an event or even just working from home, you probably want to do it with video calls or videoconferencing. And, even if your company has all kinds of teleconferencing equipment, it's probably at the office and you're probably at home. So you probably want one of the various videoconferencing programs.

You can try a lot of them free of charge. You can even use a lot of them free of charge, subject to some limitations. There are a number of options, which have varying strengths and weaknesses.

There is WhatsApp. Most people will think of it in terms of text chat, but it can make audio or video calls and even group video calls. However, it does have a four-participant limit for the video calls, and it doesn't have some of the integrated functions that others do. WhatsApp, of course, is owned by Facebook, so you may be concerned about how much of your data is being sold, although the content of the calls themselves is encrypted and therefore protected. Also, your WhatsApp account is tied to your cell phone number. There is a way to make it work with a computer, laptop, or tablet, but it's cumbersome, and while you are using WhatsApp on your computer, you probably can't use it on your phone.

Oddly, Snapchat, for video calls, does have a higher limit of 16 if you are willing to make the sacrifice of your security professional's soul. Also, I am reliably informed by a Certified Young Person that the quality of Snapchat videoconferencing is appalling.

Google, of course, has a foot in the videoconferencing door with Duo and Hangouts. The Vancouver Security Special Interest Group (SIG) used to Webcast its meetings with Hangouts, which has since fallen out of favour with our video team. We did do a test with Meet but the results were mixed (details in a bit).

At the moment, the big name in videoconferencing is Zoom. Zoom is intended for business, and the company intends that your enterprise buy a commercial account. But you can create your own account, try the system, and use it, all for free. The application just doesn't seem to be as easy to use for quick, ad hoc calls. It is definitely intended for corporate group work, and trying to establish a meeting system with non-corporate entities can be trying.

For example, there is a 'Contacts' list but that list is not easy to populate. It's not like a phone book or address book or contacts list for your cell phone or email, and it's not really easy to start off and begin to make calls. The process progresses in the manner of:

(1) Install the app.
(2) Create an account (unless you've done that already online). (You can create a Zoom account on the Zoom Website. Basically, all you need is an email address. If you've got multiple devices capable of running Zoom and multiple email addresses, it might be an idea to create multiple Zoom accounts so you can call back and forth between them to practice.)
(3) Go to the Contacts page or screen.
(4) Somewhere on the Contacts screen will be a 'plus' sign (+) (probably at or near the top of the screen). Tap the +. The screen that results might say something about search. Type in the email address (or Zoom account number, but it's probably the email address) of the person you want to add (that is, the email address they used to create a Zoom account). Then you will probably receive a message about the person/account not being found, but there will probably be a button so you can invite them. Do that and then they have to read the message and respond in order for them to be a contact.
(5) Once they accept/become a contact, they will appear on the Contacts screen (or under 'External'). Once that happens, you can click on the contact in order to obtain options to start a chat (word balloon icon) or video call (camera-looking icon).
(6) Oh, the process is not over yet. Once you start your first video call, you have to make some choices and/or tests of what audio and video you are going to use. However, on a smartphone, it shouldn't be too complicated.

(The process is somewhat easier with a commercial or corporate account. In that case, all of your co-workers are probably in your contacts already.)

Once you have established someone as a contact, calling them is a bit easier. However, Zoom does seem intended primarily for planned, scheduled meetings. On a free account, you can create a videoconference with up to 100 participants (which seems a bit of overkill). The Zoom client app does have various ways to manage this on your screen (muting your microphone, turning off your camera, hiding various participants, having a gallery view, or selecting the current speaker only). The free account is also limited to a call of no more than 40 minutes if you have more than three participants.

If you have tried out Zoom and found it difficult, what I am trying to say is that the challenging nature is not necessarily your fault. Videoconferencing really is demanding.

As previously noted, at one point the Security SIG did a bit of a 'stress test' on Google Meet as part of the attempt to create a worldwide security conference (while we're all at home and not going to onsite ones). Google Meet doesn't seem to have quite the same video quality as Zoom, and there are definitely fewer features in Meet. However, the free version of Meet has fewer limitations than the free version of Zoom. One interesting bug report: I was the 'host' of the test meeting and, at one point, the controls (ability to turn my mike on and off, ability to bring up the chat window, etc.) on my window stopped working. The meeting didn't freeze: I could see all the video going on and the various participants, and I could hear the discussion. After a while, the controls came back.

However, later in the meeting, my control froze again, and after a few minutes, the video of the participants froze but I could still hear the audio discussion. After a few more minutes, the audio cut out. After about another minute, the video suddenly started to run at 'fast forward', as if it were trying to catch up, and then I was dumped out of the conference. (My computer was fine: only the meeting window was affected.) Since I was the 'host', I wondered what that would do to the meeting, but when I rejoined the meeting, everyone was carrying on as before, so the meeting wasn't lost when I left.

A few general comments on availability and accessibility between Zoom and Meet.

Setting up contacts with Zoom is an arcane mystery. I've poked and prodded at it, and I think I can make it work, but a lot of people with a lot of experience with Zoom seem to find it difficult. I can't see much chance of newcomers doing it effectively.

You can add anyone to a Meet meeting. Anyone, that is, who has a Gmail or other Google account. (This seems to include Google-registered domain names, and the limitation is for the free version of Meet.)

You can download Zoom from Zoom.us. There are lots of people who are willing to send you a download of 'Zoom'. Almost all of these are malware. Searching for Zoom in the Apple App Store, Google Play, or the Microsoft

Store provides you lots of listings that aren't Zoom and sometimes nothing that is.

The default for Zoom meetings is now to use a password. Use of a password allows you another layer of protection against the infamous practice of 'Zoombombing', whereby people unconnected with your meeting or group can 'join' a meeting and then call out offensive messages or 'share' screens of offensive material. Most of the time, when Zoom generates a URL for a meeting, the URL includes the password for the meeting. But not always. This can become tricky for those who don't read the entire meeting invitation or announcement. Google's default is not to use a password, and I don't know whether one can be mandated when you create a meeting.

Zoom supposedly has provision for attending a Zoom meeting just using your browser without installing the Zoom application. But it does attempt to download and install software on your computer. (We'll look into details on that in a bit.)

Meet just uses your browser. No additional download necessary. However, trials using the Safari browser under Mac OS and Firefox under Windows each demonstrated significant problems. Using Chrome under both Mac and Windows worked fine. An oddity of Meet is that recently it has been very closely tied into the Google Calendar. If you create an entry in your calendar and add any guests, the calendar will schedule a Meet by default. You actually have to turn it off in order **not** to have a Meet scheduled. This can easily create confusion if you are scheduling a meeting with Zoom or some other system.

Zoom, as software, is persistent.

The Vancouver Security SIG used Zoom for a virtual/remote meeting. As part of our testing, we tried to figure out whether you actually need a Zoom account or have Zoom installed to 'attend'. The answer seems to be 'no' but with some caveats. Zoom seems to install very easily, persistently, and with minimal user involvement and maximum access. But our test results were complicated, so I'm including some screenshots to illustrate what I'm talking about. And I'm not sure that I am able to test a complete 'no-install' situation since my machines all appear to be contaminated with Zoom.

On my main desktop, I have never installed Zoom (since I don't have a Webcam on it), but I have done some work on one Zoom account (via a safe browser) and have used the Chrome browser with a Zoom install on the same account. I seldom use Edge, so I don't think I have anything installed on the Edge browser, but the install via Chrome seems to have 'contaminated' my desktop Win10 machine in its entirety.

The host set up a test meeting, and on the Edge browser, I entered the URL and got the following screen:

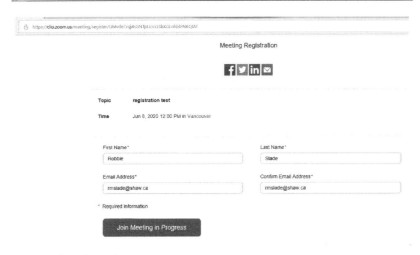

Zoom no-install test 1.

I registered using my main email address, which I have never before used for a Zoom meeting. This resulted in:

Zoom no-install test 2.

I clicked on the link (the lo-o-o-ng URL) provided. This is what I got:

Zoom no-install test 3.

At this point, I have to strongly note that I did **not** click on the 'download & run Zoom' link. I **did** click on the 'join from your browser' link. This, unfortunately, brought up:

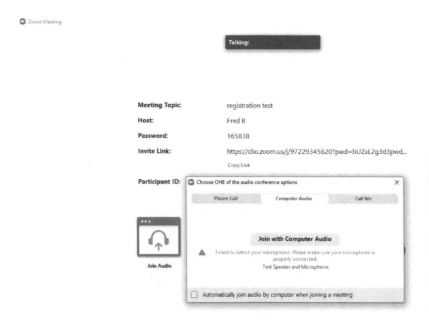

Zoom no-install test 4.

which was definitely not in my browser. Any of them. It was Zoom.

At this point, looking back at my browser, I took a screenshot of:

The meeting requires registration to join and it is not supported to join from the browser. Please try to launch meeting if you have Zoom client installed. Otherwise, download Zoom.

Zoom no-install test 5.

There are two things to note. The first is that I definitely did not click on the 'download Zoom' link. The second is to note the bottom message on the screen about Zoom_5a9bf98c8b046661.exe. I did not either run or save it. As previously noted, the fact that Zoom came up was from a previous install via a different browser.

So Zoom is very 'helpful' about installing onto your machine. As a malware researcher, I'm not sure how I feel about that. On the one hand, you can probably offer the meeting to anyone with a browser, regardless of whether they have a Zoom account or have ever used Zoom. (This does seem to be contradicted by another friend, who tried, rather desperately, to join a Zoom meeting but without any success.) On the other hand, as a drive-by download, it works great, and I'm not really thrilled about having stuff installed on my machine with lots of access that I never gave it.

Is Zoom conferencing safe to use or not?

Zoom, as a company, appears to have concentrated on ease of use, ease of installation, and functionality to the almost total exclusion of security. Up until now. As soon as a whole bunch of people started using it, a whole bunch of security researchers started noting various security weaknesses. There was the aforementioned Zoombombing. There was the zealous overemphasis on 'selling' the security features of Zoom, which had significant weaknesses. For example, Zoom calls were said to be encrypted. The encryption algorithm used was AES (Advanced Encryption Standard), which is an accepted and highly regarded function. However, it does allow you to use keys of differing lengths, and Zoom used 128 bits, which is very much weaker than the usual standard of 256 bits. In addition, Zoom calls were 'link'-encrypted, which means that Zoom would have been able to read any of the traffic since all of it went through Zoom servers. Those concerned with privacy are much more interested in 'end-to-end' encryption (sometimes referred to as E2EE), which means that only those who are actually parties to the call can access the content.

Then there was China. Zoom is ostensibly an American company. However, more than 80% of the product research and development goes on in China. And, in some early tests, in calls that had no parties in China, the meeting traffic ended up going through servers in China. More recently, an activist who had been using Zoom meetings for planning a protest to mark the anniversary of the Tiananmen Square massacre had his Zoom account deactivated.

Zoom has said that it is not owned by the Chinese Communist Party. Mind you, the same thing is said about pretty much all the companies that are owned by the Chinese Communist Party. Zoom has said that the traffic which routed through China was due to an error in server configuration, and, I note, no recent analysis has demonstrated the same routing. Zoom has promised that all 'verified' Zoom accounts will be able to use stronger and end-to-end encryption. And the activist did have his account reinstated.

However, I think it would be a good idea, in national security terms, for everyone, as much as possible, to use Zoom for inconsequential family parties and so forth so that important business and government meetings might be lost in the chaff.

At which point, I have to tell you about the dogs.

Number One Daughter has recently moved. Beautiful place, beautiful view, built on a slope, **way** too many stairs for us to live there. For complicated reasons, she now has two dogs: Marley, who is very old, shouldn't have lived anywhere near this long and has arthritis among other things, and Fera, who is young, high-strung, somewhat nervous, and **extremely** high-energy. Despite disparate ages and temperaments, 'the pack' has a great relationship.

Marley doesn't do stairs any better than I do. She has developed a weird front-legs-together-back-legs-together double-bounce method of travelling **down** stairs but can't climb back up. But when she gets to the bottom of the house, she can exit out the deck, then down more stairs, then up the slope (which is steep but at least not stairs) to the front door.

The house has an alert system for the front door. A motion sensor triggers a camera and sends a picture to Number One Daughter's cell phone. Of course, it does this over the Internet, so that she can see it when she's away from the house, as well.

However, most of the time this high-tech stuff is unnecessary. Fera, noting that Marley is at the front door, will run down through the house, out the back, onto the deck, down the stairs, up the slope, check with Marley, then race back down the slope, and up through the house until she finds Number One Daughter. Aside from the specificity of this activity, it's easy to tell that this is about Marley, because Fera has a very distinct look on her face when doing it. Fera's alert usually comes before the high-tech door system. As a matter of fact, Fera can usually make two or three round trips before the high-tech alert comes through. (Isn't it marvellous how modern technology is so much faster than the old ways we used to do things? No, I don't think so, either.)

Thing is, this is one of those cheap 'security' systems that have wretched security themselves. You know, static passwords common to all of that model and easily hackable firmware that allows someone to do interesting things with your box (and possibly your network). And it's probably sending data back to China. Maybe just metadata, although it's usually impossible to tell since it's all encrypted. But why does someone in China need to know what's going on at Number One Daughter's house? (Even though I'm using this ridiculous Charlie Chan nomenclature to keep from naming Number One Daughter.) This means that, somewhere in some monitoring station in China, someone keeps receiving and having to pay attention to alerts about Marley needing access at the front door. And Fera coming to check on her. And again. And again. And the system behind it has to commit bandwidth, storage, and processing for it.

I have **absolutely** no sympathy for those people at all.

COVID PHISHING

I got the following message via text/SMS on my phone:

'(Notification—ALERT) Dear client, Scotiabank is working with the Government to make the Emergency COVID-19 Benefits deposits easier. To complete your Benefit demand. Please visit now: www.Scotia-Online-COVID19.com'.

I do have an account with Scotia, but I have not had texts from Scotia and I do not recognize the domain. In addition, the message seems to have some errors in terms of grammar and general English usage. So, being a Professional Paranoiac of the first water, I did a domain lookup.

scotia-online-covid19 doesn't exist.

But, being as paranoid as I am, I noticed that O (capital letter 'o') and 0 (digit zero) look the same in most screen fonts.

And, voilà! scotia-0nline-c0vid19.com (with zeroes in 0nline and c0vid19) **does** exist! It's registered in Panama. A lot of companies with questionable business models are registered in Panama.

(I'd advise not going anywhere near it.)

This one was directly targeted at my bank account, but there are huge numbers of scams that spring up with every disaster. As noted previously, some pretend to be charities, some will target relief programs, and many are just more of the same in terms of the regular ongoing run of spam and scams. But a number of the frauds do seem to be specific to the pandemic.

There are the puppy scams. With everyone staying at home, huge numbers of lonely people are suddenly wanting a pet. (Making such a major, ongoing, and life-modifying decision at a time of crisis and high stress is perhaps not the best of ideas, but we covered that in business continuity planning.) So the scammers, noting the exploding market, are happy to sell you puppies! By and large, of course, puppies that don't exist. And, of course, you will be paying by credit card. Which they'll steal.

Which brings us to the shipping phish. You will probably have ordered something online during the pandemic since stores were closed and everyone was staying home. (Amazon and eBay and everyone who has any kind of online presence have been doing land-office business.) So you're used to receiving notices of items shipping. And you might have received a few emails claiming to be from a courier company, asking you to pay delivery charges before they can deliver a parcel that is addressed to you. That seems possibly reasonable, and the charges are usually very small.

Now, there are two items to note here. One is that millions of very small charges still add up to a lot of cash. The other is that, when you pay that very small charge with your credit card, you are providing all the information that the scammers need in order to make other, much larger charges on your credit card.

COVID-19, 'SHAMING', AND CYBERBULLYING

Watching the news recently, I saw two different government officials, from different provinces, and one television commentator recommend that, if you saw any evidence of hoarding, you should 'shame' the perps on social media.

I can understand the frustration. I really can. Hoarding is insidious and is terribly destructive to the community as a whole during a crisis. Those who are 'bulk buying' in order to resell at a huge mark-up are despicable. (I don't care if you think it's the 'American Way'. It's despicable.) Those who are stealing hospital supplies should definitely be prosecuted.

But I don't think those who recommend public social media shaming understand social media. 'Shaming' is one post away from cyberbullying. If social media is about anything, it's about disproportionate overreaction. It has no off switch, no censorship or moderation, and no 'right to be forgotten' or means of redress. We've already seen instances of mistaken or misunderstood postings going wildly viral during this crisis, and we're going to see more. Don't add to them.

This bit possibly should have been up with the section on integrity because many of the calls for shaming are, in fact, based on misinformation or even simple misunderstandings. As Dr. Bonnie said in response to pretty much every reporter's question about people being in the wrong place or doing the wrong thing, you may not know the whole story.

As an example, at one point during the lockdown, a number of groups (mostly on the right of the political spectrum and sometimes slightly to the right of Attila the Hun) were holding protests calling on people to 'liberate' and 'reopen' various states. Some had Websites in order to organize the protests. People researching the phenomenon found that a number of domains with similar patterns had been registered by one person, and they started a campaign to attack that person on social media.

The problem was that that person had no connection with the protests. He later described himself as an aging hippie and, if anything, somewhat left-leaning politically. He had, in fact, noticed exactly the same pattern of Websites and domains and had registered a large number of those domains so that the protest groups could not use them. At considerable cost to himself. Financially, initially, but later in terms of the attacks made against him by those who also opposed the protests. (No good deed goes unpunished.)

Then there was the Starbucks barista. Serving a customer, he noted that she wasn't wearing a face mask. Before he could even point out the company policy about face masks, she (a) took offence, (b) took his picture, and (c) posted it on Facebook, along with his name, expecting everyone to go to the Starbucks and insult him. It didn't quite work out the way she thought. Tons of people expressed support for him online, and one started a GoFundMe page for 'tips' for him. He got at least $80,000 in tips.

Online shaming is really hard to do just right.

Law, Investigation, and Ethics

11

OVERREACH

'They that can give up essential liberty to obtain a little temporary safety deserve neither liberty nor safety'.

– Benjamin Franklin, *Historical Review of Pennsylvania* (1759)

Many jurisdictions are using the CoVID-19 crisis for their own purposes. Governments are using contact-tracing applications and systems to conduct surveillance on their citizens or using regulations about isolation and distancing to forbid protest and stifle dissent. A number of states in the US are determining that abortion clinics, alone of all medical facilities, are non-essential and should be shut down. We must be careful, in security, that we are putting policies in place for real reasons of security and not for other purposes. (In terms of ethics, we must use our powers only for good, never for evil.)

This isn't necessarily simple, in terms of the pandemic. Like most situations (but more explicitly than most situations), CoVID-19 points out that we have to balance competing legal principles and carefully examine the intent of the original constitutional documents. Americans would say that they are promised life, liberty, and the pursuit of happiness. Don't they have the right of liberty to conduct their businesses without being restricted by lockdown orders? Don't they have the right to pursue happiness by going to nightclubs? Ah yes, but 'life' comes first.

PRIVACY

Quite apart from contact tracing, any number of privacy issues are involved with CoVID-19. For one thing, it's a medical issue, and medical information seems to be the second (how much you are paid is the first) highest privacy concern people have. It's quite a legitimate concern in many cases: being diagnosed with certain diseases can lose you a job or insurance coverage, and in all too many cases, you become shunned by the community, often for something that is not your fault in any way.

EVIDENCE

One of the rules of evidence is that, in order to present it in court, it must be relevant to the case. The importance of this has been illustrated by the different types of testing and various pieces of research that have been conducted and published while the pandemic has been raging.

For example, RNA testing has revealed that SARS-CoV-2 has been detected in feces. Does this mean that sewage is an infection risk? (Look, work with me here. Sewage is always noxious stuff, and you can become infected with a number of nasty diseases from it, quite apart from SARS-CoV-2.) Well, we don't know. We know that the RNA tests detect sections of the virus code, but they don't necessarily indicate that the virus, even when detected, is viable. Possibly more study is warranted, but at the moment, no other evidence indicates that fecal contamination is a major vector of the virus. (The fact that we can detect the virus in sewage may provide us with a tool for finding out how prevalent the virus is in a given community, but we need more study on that too.)

In the same way, we have studies indicating that the virus can be detected (via RNA tests) in the air for hours after an infected person has left. Does that mean that SARS-CoV-2 is an airborne virus and that transmission can take place by these particles in the air? Well, again, not necessarily. For the same reasons given above, we may be detecting non-viable traces of the virus, so we don't yet really know what danger there is in aerosol transmission. So far, the evidence of actual transmission seems to indicate that the risk of infection is low. The fact that we can use RNA tests to detect viral fragments in the air is not really relevant to our risk of infection.

RACISM

I don't know whether it properly belongs with law or ethics, but it certainly belongs in this domain.

One thing that has horrified me, possibly above all the other repellent issues in this awful crisis, is the extreme and rapid rise of overt and blatant racist physical attacks and misbehaviour. It has been seen in many ways: verbal harassment on the street, physical attacks on transit, racist graffiti, all the way up to shunning migrants and the closing of borders where no real need to do so exists.

It is possibly because I am Canadian that I find it so deeply disturbing. Canada, unfortunately, is not immune to racism, but we do have, as a national characteristic, tolerance-to-the-point-of-vice. Much more than the United States, we are a cosmopolitan and polyglot society. Whereas Americans talk about the 'melting pot', where everyone comes to the US and becomes American, Canada talks of the cultural mosaic, where everyone comes to Canada and contributes something from their point of origin. There is a national joke that the definition of a Canadian is a DP with seniority. For non-Canadians, I have to explain that DP stands for 'displaced person', an old term (dating from the two World Wars) for refugee, but also that DP is itself a racist term. Thus, the joke is not just about the demographic composition of the country (every Canadian is a DP: even the First Nations peoples came here from somewhere else, most likely Siberia) but is also a sly dig at racism itself (and a slightly more overt self-mockery of Canadians who may think they are above or immune to racism). (Canadian jokes do tend to be complex and layered. Sorry.)

The George Floyd/Black Lives Matter protests may relate to whatever cultural forces are behind the racism. As previously noted, racial bias in law enforcement is nothing new and is widely known and acknowledged, though mostly without being effectively addressed. There have been other deaths and other protests. Colin Kaepernick had his own, completely non-violent, protest for many years. It is, of course, entirely possible that much of the success of the movement has simply been because people were fed up with having to stay at home during the pandemic and badly wanted an excuse to go outside and be with people. But it also seems that society appears to have been ready, at this point, to take a more serious look at the issue. It will not be fixed overnight, but it would be heartening if some meaningful change came out of this terrible mess.

Be Kind. Be Calm. Be Safe

12

BE KIND

Dr. Bonnie Henry, provincial health officer for the province of British Columbia (BC), in Canada, was, on an almost daily basis, for months on end, the voice (and, to be fair, the architect) of the government's response to the coronavirus crisis. Almost every day at 3 PM, she would hold a virtual press conference with reporters calling in by phone to submit questions. At this point, I do not want to dwell on Dr. Bonnie's qualifications, background, or the excellence of her guidance. I do want to say that these press conferences amply demonstrated that Dr. Bonnie is the person for whom the phrase 'patience of a saint' was meant to describe.

Day after day, reporters asked dumb questions, questions with no answers, questions that had been amply answered, and trick questions meant to elicit some sensational sound bite. Day after day, Dr. Bonnie disappointed the journalists of the sensational and reassured the rest of us. She answered the dumb, impossible, and repeated questions, and I never, ever saw her lose her temper, even when I, at home, was losing mine, at their 17th attempt to obtain a 'one size fits all' quote that could later be used to attack Dr. Henry and boost newspaper sales.

One of the most common questions that reporters would endlessly repeat was some variation on 'How viciously should we attack people who don't keep six feet away from other people in line or pass a hand-sanitizer dispenser without gooping their hands or go to the beach when I don't feel safe doing so?' Dr. Bonnie's constant answer was the first part of her triple mantra: Be kind.

She usually backed it up with a rationale for not taking instant offence at every perceived infraction. For one thing, you don't know everything. That group of a dozen people in the park? Maybe they are all members of one family and household. They are going outside and breathing some fresh air. They are carefully keeping distanced from everyone who is not part of their household.

They aren't actually breaking any rules. That person not wearing a face mask? She has chronic obstructive pulmonary disease, and a mask is almost a death sentence for her. That person wearing a face mask? He doesn't have CoVID-19 but does have a cold and needs groceries but wants to minimize the risk of spreading his cold to anyone else.

I do understand some of the journalists and some of their motivation. We are all under enormous stress. We are possibly in danger of our lives. We are certainly not living as we would wish: life, as we know it, has been indefinitely suspended. And that 'indefinitely' only adds to the stress. We don't know when this will end. And we want someone to blame. What better candidates than those fools who are doing what we don't approve of? Well, Aleksandr Solzhenitsyn observed very much the same thing and noted the same reaction from the general public. In *The Gulag Archipelago* (Harper & Row, 1973), he noted: 'If only it were all so simple! If only there were evil people somewhere insidiously committing evil deeds, and it were necessary only to separate them from the rest of us and destroy them. But the line dividing good and evil cuts through the heart of every human being. And who is willing to destroy a piece of his own heart?'

There is a wonderful principle for when you feel like people are acting specifically in such a way as to cause you harm. It is called Hanlon's razor and it is formulated along the lines of 'Never attribute to malice that which can be adequately explained by stupidity'. In other words, stop, take a breath, and consider that maybe they aren't out to 'get' you. Maybe they don't know what you know. Maybe they aren't as smart as you. Maybe, from their perspective, they are doing the best they can, even if that best isn't perfect. Maybe, instead of going on the attack against them, it might be more effective to think about how you might help them make better decisions and take better actions.

How nice, I hear you say. But what does this teach us about security? Well, quite a lot, when you think about it. Even before we delve into personnel management issues, think about one thing. Why are we doing security?

We aren't doing security just because we know the technology. (Hopefully we have this background.) We aren't doing security just because it's fun. I think it is: a lot of people disagree. One of them is my friend Ron, who was an absolutely stalwart support for the Vancouver Security Special Interest Group (SIG) for over three decades. He worked hard for us, he did lots of the unregarded admin jobs that nobody wanted to perform, and he was always reliable. When he retired from his day job in security, I asked if he was still going to come out to the SIG meetings. He said that he wouldn't: he had never been all that interested in security and just did it because it was his job.

While that is not a great example for career planning, it is an absolutely perfect answer to the question of why we are doing, or should do, security. Ron did security (and very well too) to support his enterprise. We are doing security to support our respective businesses, employers, and fellow workers.

We are doing it to be kind to those around us and to keep our colleagues safe and sometimes to keep our communities safe as well.

Looking at it from this perspective can be extremely important in terms of your overall attitude to security. We, as security professionals, are often seen as impediments: we are 'the knights who say no'. We are the people who are always telling others what they can't do. And we, all too often, see ourselves that way: we are the ones who know the risks, and we have to keep these fools from hurting themselves.

But, when you turn it around—we actually are here to help—it can make a big difference. There is an actual reason we say no, and it isn't just to impede things, so maybe we don't have to say no. Maybe we can say 'yes, but'. So when the next C-suite executive comes up with the next great idea he got from an in-flight magazine to implement blockchain on all your logins, you can tell him that blockchain is a fascinating field of study and you'd dearly love to help implement it, but it does take a lot to set it up correctly, so you assume he'll be supporting your ask for $500 K for the initial implementation and the extra two full-time equivalent (FTE) positions to manage and maintain it.

OK, perhaps that is a bad example—you know he won't, and you'll never hear about it again, so that's equivalent to saying no. So, maybe, when you find that lots of the staff are choosing really stupid passwords, you can start your security awareness sessions on password choice by saying that you know the staff want to perform their tasks, and their system accounts are there to help them in their jobs. But if they choose easy-to-guess passwords and their accounts are hacked, the company has to take their accounts (and therefore their jobs) away until things can be fixed, and you have some pointers that help to ensure their accounts are safer. You are there to help. To be kind.

BE CALM

On May 22, 2020, Dr. Bonnie Henry was 'adopted' by the Gitxsan First Nation and given a Gitxsan name. The province's top doctor was named 'Gyatsit sa ap dii'm', meaning 'one who is calm among us'.

It is well known that, when making a business continuity plan, you always try to be as detailed as possible so that the people following the plan do not have to make decisions while they are stressed and at their worst.

Be calm. Take a breath. Relax. Change the things you must change, let go of the things you can't, and know the difference.

Once again, many reporters had a constant theme in all the question periods following Dr. Henry's press briefings. Whom do we blame? Whom can we punish? Who is at fault? Whom do we call out? Whom can we rail about in our editorials? The reasons they want to blame someone vary. Why is there a shortage of PPE (personal protective equipment)? Why are people going to the vacation house? Why are people congregating at the beach? Why are protesters showing up in large crowds? Why don't people use masks? Why are people relying on masks? Why aren't you making stronger rules about masks? But mostly they all want someone to scream at.

I can understand the feeling. We are faced with a big, scary, complex, unknown risk on top of risk on top of more risks. Everyone is frightened. Everyone is tense. Everyone is short on sleep and having bad dreams and feeling helpless. It's quite clear why everyone wants someone to blame for this whole mess or even just a little part of it. It's also completely useless. Trying to assign blame in the middle of a disaster is one of the worst decisions you can make. And also the least productive.

BE SAFE

As previously noted, the primary rule in emergency management is 'First, do no harm—to yourself'. If you do not take care of yourself first, you can't help others and you risk becoming part of the problem. Especially in a pandemic, as with the computer virus example I gave earlier, if you become infected, you might require medical resources that are in short supply and high demand, and you become a source of risk to everyone around you, most particularly to those closest to you.

Firefighters are brave: that is not in question. But, movies and TV shows notwithstanding, they do not just run pell-mell into a burning building to save someone. They have trained and practiced. They have protective equipment and spend hours making sure that equipment is maintained and ready to use at need. When they do run into a burning building to save someone, it is with a plan of how they will run back out again. (A young friend, who was, at the time, interested in becoming a firefighter, once ran into a burning house to put out the fire. Since the house was, as they say, 'fully involved' and he was armed with a one-pound dry-chemical fire extinguisher, I had to risk my life by running in after him to haul him out.) (OK, it was necessary in that instance, but I do not recommend it.)

Another factor to note: many people are starting to show signs of 'quarantine fatigue'. This is a special case of 'caution fatigue', a well-known psychological phenomenon that is a result of the way that humans react to threats and stress. We are wired to react to changes, and when a threat is prolonged, we start to become acclimated to it and no longer regard it as a threat. Therefore, as the pandemic goes on, people start to relax their following of the rules that they willingly followed at the beginning of the crisis. Don't.

We can't give up on the rules until we develop (and test) an effective vaccine and have enough doses of it for almost everyone. Even though various restrictions may be dragging on and even though case numbers may be going down, follow the rules and stick with the World Health Organization's Five Heroic Acts.

Will We Win?   13

'Fairy tales are more than true – not because they tell us dragons exist,
but because they tell us dragons can be beaten'.

– G. K. Chesterton

Will we win out over the virus? This is actually a very interesting question in
regard to security since it touches on an extremely complex risk management
calculation with an enormous number of variables and an almost infinite number
of possible outcomes.

Human beings are not very good at risk analysis of the types most common in our complex modern world. We have all kinds of cognitive biases that
were shortcuts to decisions in the days when all we had to worry about were
sabre-toothed tigers in the tall grass but that are a positive nuisance these days.
We tend to anchoring, which is our tendency to build a mental model around
as few as one or two examples, which is then backed up by confirmation bias,
which tends to make us pay attention to later examples that support our model
and disregard those that contradict it. There is automation bias, which prefers
any kind of process for decision-making to rational analysis and tends to make
us trust machines rather than ourselves. There is the ambiguity effect, which
makes us so uncomfortable with uncertainty that we will jump on the first
answer we find (and then cling to it with all the force of anchoring and confirmation bias behind it). These are only a few of the problems with our thinking
and we have to guard against them.

To begin with, in our analysis of whether CoVID-19 might beat us,
there is Dr. Adam Lawton's since-deleted March 28, 2020 post on Twitter
(@DrAdamLawton) to the effect that if you are a new single-stranded RNA
virus and want to survive in a dangerous universe, attacking the only DNA-based organism that can sequence your genome and that has historically eradicated more species than any other living thing is probably a bad idea. But,
while it is possibly a comforting thought momentarily, there are some flaws in
the logic of relying on the fact that humans, as a species, are pretty mean. There
are a number of positive possibilities, though, and some encouraging signs.

Absent the world leaders in CoVID-19 cases (currently the US, Brazil, Russia, and India), most places in the world either have relatively low numbers of cases or have passed through the first wave. While every death is a loss and businesses of all types have suffered serious damage, lockdowns and the basic hygiene rules of the World Health Organization's Five Heroic Acts do seem to be effective in containing the disease. Enterprises of almost all types are adjusting in various ways: manufacturers are retooling to produce high-demand items, restaurants are going to a takeout model, and sports leagues are figuring out procedures for at least holding televised competitions. (Events planners do seem to be slow off the mark in terms of creative adjustments and alternatives, which seems rather odd.)

The curve has been flattened. We now know what we need to do to slow the spread of the disease to manageable levels, even if we never cure it.

Of course, we actually have to do the things we know we should be doing. As I am finishing up the text and last edits of this book, there is great concern over the number of very large private parties that are going on in contravention of everything we know about how to avoid infection. Maybe we don't deserve to survive, as a species, if we can't go a couple of years without mass concerts, alcohol, huge parties, and completely thoughtless behaviour. You want to get together with a whole bunch of random strangers? Too bad. I want to hug my grandchildren, go to conferences, and go to restaurants and sit there and eat hot food without having to eat it cold after taking it all the way home. Can't. This is an emergency and we have to behave appropriately.

(Have I mentioned the memes harkening back to the two World Wars and similar events? The memes point out that those people had to risk their lives to save humanity and all you have to do to save the world is stay home and watch TV.)

You're tired of all the rules? Well, we could probably reduce them to just one. We'll probably survive if we give up alcohol for a couple of years. Look at the facts. Bars have to be closed and restricted. Bars serve alcohol. Restaurants have new rules imposed. Restaurants serve alcohol. Concerts are banned. Concerts involve alcohol (even when they aren't supposed to). We can't go to sports arenas. Sports arenas serve alcohol. Parties are dangerous. Parties usually involve alcohol. (Let's face it: every party except birthdays for ten-year-olds has alcohol, and even there the parents are probably drinking.) Why aren't the conspiracy theorists on to this? There is much more evidence that alcohol causes coronavirus than that 5G does, and the tin-foil-hat crowd is out there burning down cell towers.

We haven't yet found a cure. Hydroxychloroquine seems to be useless and even dangerous (unless very carefully managed). Oleandrin comes from a poisonous plant. (Lots of drugs do, but there isn't really any evidence yet that oleandrin has any medical benefits.) Remdesivir does seem to slightly reduce time sick and slightly reduce the possibility of death. Dexamethasone seems to be showing some promise, but we don't know much about the benefits (and risks) yet (as of this writing). But we've been at this less than a year. And, if it is going to take a while, we know what we have to do to buy time. Follow the rules.

We haven't got a vaccine. Yet. We have never successfully created and tested an effective vaccine for this class of viruses. We do have more than 100 candidates, and we have a number of approaches that have led to vaccines for a wide variety of viruses in the past. And we know what we have to do to buy time for the development and testing.

There is a recent book on climate change, *The Uninhabitable Earth* (Tim Duggan Books, 2019), by David Wallace-Wells. He notes that the realization that global climate change is caused by humans should give us comfort, not despair. We are responsible for the change and therefore we are in command and can change the situation. By similar logic, in terms of the pandemic, we do know how to beat it. We just have to follow the rules.

A couple of days ago, I gave the first actual presentation of the talk on which this book is based. There was one question that I found rather odd, and I'm not sure that I understood what motivated it, but someone asked whether, in view of the various research being done, effort was being wasted on novel research, that could be put into reproducing the results of earlier research, which is the acid test for scientific research. As I said, I'm not sure what prompted the question, but I have to assume it is based on a fear that somehow the research is being distracted by new questions and should be finishing the existing work before moving on. But science is not a single effort. Hundreds of thousands of people around the world are looking at what has been done, what is being done, and what could be done. If you think that isn't enough, maybe you should encourage your children to pay closer attention to science classes in school and get into research when they grow up. But I don't think we need worry that existing research will not be tested and challenged. Jacques Plante, the great ice hockey goalie, once asked: 'How would you like to have a job where, when you make a mistake, a big red light comes on and 15,000 people boo?' Science is much, much worse, in that regard, than hockey. In science, when you do something right, all of your friends try to prove you wrong.

Forty years ago, I read Aleksandr Solzhenitsyn's *The Gulag Archipelago* (Harper & Row, 1973). Some of the passages have stayed with me ever since. Like this one:

> What about the main thing in life, all its riddles? If you want, I'll spell it out for you right now. Do not pursue what is illusory—property and posi-tion: all that is gained at the expense of your nerves decade after decade, and is confiscated in one fell night. Live with a steady superiority over

life—don't be afraid of misfortune, and do not yearn after happiness; it is, after all, all the same: the bitter doesn't last forever, and the sweet never fills the cup to overflowing. It is enough if you don't freeze in the cold, and if thirst and hunger don't claw at your insides. If your back isn't broken, if your feet can walk, if both arms can bend, if both eyes see, and if both ears hear, then whom should you envy? And why? Our envy of others devours us most of all. Rub your eyes and purify your heart—and prize above all else in the world those who love you and who wish you well. Do not hurt them or scold them, and never part from any of them in anger; after all, you simply do not know: it might be your last act before your arrest, and that will be how you are imprinted in their memory!

Here, perhaps, is the final security lesson. Concentrate on what is important. Security, like life, is full of trivial details. Do not get caught up in non-essentials just because they are immediately in front of you.

Index

A

ACE2, 10
Anderson, Ross, xvii
antibodies, 8, 33, 34, 58
    in testing, 60–61
assurance requirement, 37–38
asymptomatic, 12, 31, 38
availability, xvi, 1–3, 14, 55, 83
    and confidentiality, 3

B

basketball, xii
BCP, 41–47
beacon, 5–6, 55–56
Bluetooth, 5–7, 55–56

C

capital risk, 42–43, 45
CISSP, 30, 47
clickbait, 13
computer security, xiv, 1, 64
computer virus, 7, 11, 57–60, 63, 100
confidentiality, xvi, 1–3, 6, 55
contact tracing, 3–7, 37–38, 55–56, 94
controls, 31, 38, 73–74, 83
coronavirus, xvi, 10, 11, 15, 61
    crisis, xiii, 79, 97
    description, 8, 10, 59–60
CoVID-19, xi, xvi, 8, 10
    bubble, 73
    genetic material, 58
    and interferon, 59
    rates, 20
    risk 25, 64
    and testing, 60, 76
cyberbullying, 90–91
cybersecurity, 2

D

data security, 1
de-anonymization, 3

defence in depth, 31
detection, 33, 58–59, 80
disinformation, xiii–xiv, 11, 14
Dix, Adrian, xii, 33, 68
DNA, 58, 103
DP-3T, 56
droplet, 11–12, 33, 59–60, 62
    and masks, 20–22, 70–71

E

efficiency, 5
emergency management, xiii, xvi, 15, 25,
    47–48, 100
errors, xiii, 6–9, 31, 73
    testing 33–34, 61

F

false negative, 7
false positive, 6–8
fear, 11, 13, 69
forfend, 44
fraud, 13–14, 19, 22, 65, 90
front-line medical workers, xiii–xiv
functional requirement, 37–38

G

Google Meet, 83

H

hand sanitizer, xi, xiii, 20, 32
handwashing, 13, 20, 22–23, 32, 71
    as a control, 38
    effect on virus, 60
Hanlon's Razor, 98
Henry, Bonnie, xii, 65
    and the press, 19, 68, 69
    kind calm safe, xv, 7, 95, 97
hoarding, xiii, 14–15, 64, 90–91
hotspot, xi

I

infection, 3–4, 10, 36, 55–56, 58
   detection, 33, 61
   prevention, 70–71, 73
   rate, 8, 26
information security, xvi, 2, 42, 74
Infosec, 2
integrity, 1–2, 6, 10, 55, 91
Interferon, 59
(ISC)², xvi, 2, 17

L

lockdown, 20, 29, 49, 72, 104
   protests, 91, 93

M

mask, 12, 20–24, 45, 68–72, 91
   and face shields, 39
meet, 81, 83–84
mental health, xii
misinformation, 9–11, 63, 91

N

N95, 12, 20, 45, 64, 70

P

pandemic, xi, xii, xv, 2, 3, 8, 12, 14, 21, 23,
      26, 38, 41, 43, 44, 52, 57, 59, 61,
      64, 68, 69, 72, 74, 90, 93–95, 100,
      101, 105
panic-buying, xi, xiii, 45, 64
physical distancing, 13, 20, 22, 23, 52, 71
positivity rate, 8
probability, 7, 12, 36, 61, 62, 70

R

recovery, 24, 41–49, 73, 74
redundant, 41, 43
requirement, 5, 35, 37–38, 47, 48, 68, 78
restoration, 49
risk, xi, xiii, xiv, 3, 7–9, 11–13, 17–23, 26–27,
      45, 55, 56, 61, 62, 64, 72, 76, 94,
      98–100, 103–105

risk management, 20, 22, 25–26, 42, 62, 103
RNA, 33, 45, 58, 60, 61, 94, 103

S

SARS-CoV-2, xi, xiii, xv, 8–11, 17, 18, 23, 26,
      59–61, 71, 79, 94
Schwartau, Winn, 36
security awareness, xvii, 63, 67, 99
security theatre, 17–19, 23, 69
serology, 8, 33, 45, 60, 61
shaming, 90–91
six feet, 7, 24, 38, 60, 62, 67, 97
social engineering, 19–20, 69, 71
Solzhenitsyn, Aleksandr, 103
supply chain, 41, 43–45, 68
surveillance, 3, 24, 55, 69, 93

T

testing, 3, 33, 37, 45, 60–64, 75–78, 84,
      94, 105
two metres, xvii, 7, 11, 12, 32, 37, 38, 52, 53,
      60, 64, 67, 72

U

ultraviolet, 18

V

ventilators, xiv, 32, 67
videoconferencing, 79–83
virus, xii–xv, 3, 7–9, 11–13, 15, 18, 20, 22, 23,
      25–27, 32–34, 38, 44, 45, 53,
      57–64, 68–71, 94, 100, 103
   wash your hands, xiv, 38, 71

W

Webinar, 79–81
Wehlou, Martin, 43

Z

Zoom, 76, 80, 82–88, *85–87*